STRUCTURAL PACKAGING

DESIGN YOUR OWN BOXES AND 3-D FORMS

Paul Jackson

D1394092

NEW COLLEGE NOTTINGHAM

250586

LAURENCE KING

Published in 2012 by
Laurence King Publishing Ltd
361–373 City Road
London EC1V 1LR
United Kingdom

email: enquiries@laurenceking.com
www.laurenceking.com
© 2012 Paul Jackson

All rights reserved. No part of this
publication may be reproduced or
transmitted in any form or by any means,
electronic or mechanical, including
photocopy, recording or any information
storage and retrieval system, without
prior permission in writing from
the publisher.

Paul Jackson has asserted his right under
the Copyright, Designs, and Patents Act
1988, to be identified as the Author of
this Work.

A catalogue record for this book is
available from the British Library.

ISBN: 978 1 85669 753 8

Designed by Struktur Design
Box production: Gilad Dies Ltd,
Holon, Israel
Senior editor: Peter Jones
Printed in China

New College Nottingham
Learning Centres
250586   741.6

STRUCTURAL PACKAGING

DESIGN YOUR OWN BOXES AND 3-D FORMS

Paul Jackson

Laurence King Publishing

Contents

Over the past two decades or so, a steady flow of packaging source books has published many hundreds of ready-to-use templates (called 'nets') for a broad range of cartons, boxes and trays. These excellent books can be extremely useful to a reader seeking an off-the-peg solution to a design problem, but they don't describe how bespoke packaging can be created, implying that innovation is something best left to the specialist packaging engineer.

I disagree!

In the 1980s I developed a simple system – a formula, even – for creating the strongest possible one-piece net that will enclose any volumetric form which has flat faces and straight sides. In its most practical application, it is a system for creating structural packaging.

This system of package design has been taught on dozens of occasions in colleges of design throughout the UK and overseas. I have routinely seen inexperienced students create a thrilling array of designs that are innovative, beautiful and practical, some of which have gone on to win prizes in international packaging competitions. It has also been taught on many occasions to groups of design professionals, who have used it to develop new packaging forms.

This book presents that system.

However, it is more than just a system for creating innovative packaging. I have used it frequently in my own design work in projects as diverse as point-of-purchase podia, exhibition display systems, mail-shot teasers, teaching aids for school mathematics classes, large 3-D geometric sculptures, 3-D greetings cards ... and much more. It is primarily a system for creating structural packaging, but as you will see, when properly understood, it can be applied to many other areas of 3-D design.

In that sense, this is a book not only for people with an interest in structural packaging, but also for anyone with an interest in structure and form, including product designers, architects, engineers and geometricians.

01:

BEFORE YOU START

1.1 How to Use the Book

The book presents a step-by-step system to design packaging and other enclosed volumetric forms. You are strongly encouraged to read it sequentially from the first page to the last, as though it were a novel. To flick casually backwards and forwards, stopping randomly here and there to read a little text and look at a few images will probably not be enough for you to learn the method with sufficient rigour to gain any significant and lasting return from the book. Used diligently, the book will enable you to create strong, practical forms of your own design. Used superficially, it will perhaps teach you little.

Chapter 2, How to Design the Perfect Net (pages 14 to 37), is the core of the book. The chapters that follow show how the methods of net design presented in it can be applied. The final chapter presents a series of packaging forms created by students of design at the Hochschule für Gestaltung, Schwäbisch Gmünd, Germany, developed from the forms seen in previous chapters.
By working through the book sequentially, you should reach the final pages understanding enough about the theory and application of the net design method to create your own high-quality, original work.

My strong recommendation is to resist temporarily the urge to create. Instead, open yourself to learning and then to applying creatively what you have learnt.

1.2 How to Cut and Fold

1.2.1 Cutting

If you are cutting card by hand, it is important to use a quality craft knife or, better still, a scalpel. Avoid using inexpensive 'snap-off' craft knives, as they can be unstable and dangerous. The stronger, chunkier ones are more stable and much safer. However, for the same price you can buy a scalpel with a slim metal handle and a packet of replaceable blades. Scalpels are generally more manoeuvrable through the card than craft knives and are more help in creating an accurately cut line. Whichever knife you use, it is imperative to change the blade regularly.

A metal ruler or straight edge will ensure a strong, straight cut, though transparent plastic rulers are acceptable and have the added advantage that you can see the drawing beneath the ruler. Use a nifty 15cm ruler to cut short lines. Generally, when cutting, place the ruler on the drawing, so that if your blade slips away it will cut harmlessly into the waste card around the outside of the drawing.

It is advisable to invest in a self-healing cutting mat. If you cut on a sheet of thick card or wood, the surface will quickly become scored and rutted, and it will become impossible to make straight, neat cuts. Buy the biggest mat you can afford. Looked after well, it will last a decade or more.

A scalpel held in the standard position for cutting. For safety reasons, be sure to always keep your non-cutting hand topside of your cutting hand.

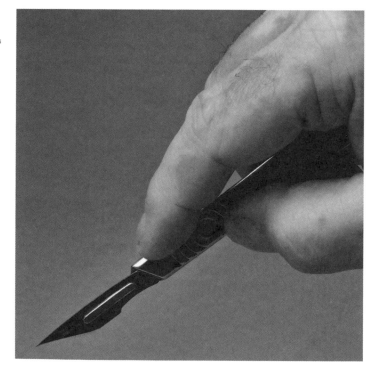

1.2.2. Folding

While cutting paper is relatively straightforward, folding is less so. Whatever method you use, the crucial element is never to cut through the card along the fold line, but to *compress* the fold line by using pressure. This is done using a tool. Whether the tool is purpose-made or improvized is a matter of personal choice and habit.

Bookbinders use a range of specialist creasing tools called bone folders. They compress the card very well, though the fold line is usually 1–2mm or so away from the edge of the ruler, so if your tolerances are small, a bone folder may be considered inaccurate.

A good improvized tool is a dry ball-point pen. The ball makes an excellent crease line, though like the bone folder, it may be a little distance away from the edge of the ruler. I have also seen people use a scissor point, a food knife, a tool usually used for smoothing down wet clay, a fingernail (!) and a nail file.

But my own preference is a dull scalpel blade (or a dull craft-knife blade). The trick is to turn the blade upside down (see below). It compresses the card along a reliably consistent line, immediately adjacent to the edge of the ruler.

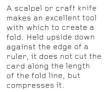

A scalpel or craft knife makes an excellent tool with which to create a fold. Held upside down against the edge of a ruler, it does not cut the card along the length of the fold line, but compresses it.

1.3 Using Software

When I teach, I must by necessity ask my group to construct their nets manually – it simply isn't practical to design with a computer. So we make nets using a hard pencil, rulers, a protractor, a pair of compasses, set squares and – of course – erasers. In truth, this is absolutely the best way to learn how to design a net. Later, when a perfect net has been designed, it can be drawn using a computer.

However, the correct ways to draw accurate squares, parallel lines, polygons and so on by hand, and how to calculate angles, are rarely taught now in schools or in design colleges, so when I teach, a lot of time is given to explaining the basic principles of technical drawing. To explain basic TD within these pages is beyond the scope of this book, so the reader wishing to construct by this manual method is encouraged to seek information elsewhere.

More likely though, the reader will use the system of net design presented in this book to create a rough net, which will then be drawn accurately on a computer.

There is a wide choice of excellent CAD software suitable for drawing nets, some of which is available in less powerful Freeware versions. It is also possible to use graphic design software, though geometric constructions can sometimes be a little laborious to make. Essentially, any software that can create two-dimensional geometric constructions is suitable. If you already have a reasonable knowledge of a particular CAD or graphics application, you can probably use it to create accurate nets. If you have no such knowledge, one of the Freeware CAD applications is a good place to start. If that is beyond you, simply purchase a basic set of inexpensive geometry equipment (the list is in the first paragraph, above) and make everything by hand.

1.4 Choosing Card

All the examples photographed for the book were made with 250gsm card. If you are making examples from the book, or creating your own maquettes, this is the recommended weight to use. If you know you will eventually use thicker boards, or even corrugated cardboard, for your final design, it is still recommended that you make maquettes in 250gsm card before moving up to the heavier weights. Try to use a matt card, rather than a coated glossy card, as a matt surface will fold better, has more grip to lock a net tightly together, can be drawn on more easily, and is generally more workable and user-friendly than coated card. If you need to impress someone with what you have made, a bright white card creates better-looking boxes than a dull white or off-white card.

If you are designing a one-off package for a personal project, or for a low handmade production run, you may choose any type of card. However, if you are intending to manufacture your design in quantity, you will need to consult a specialist packaging engineer to discuss which card is best for your needs. More about this can be found in How Do I Produce My Box? (see page 126).

One more thing: although the book features packaging made in card, many of the nets can be adapted to plastic or, more specifically, polypropylene. The possibilities of creating in polypropylene are immense and visually exciting, especially if the material chosen is translucent or transparent.

1.5 Glossary

Like most specialist activities, structural packaging has a terminology all its own, though many of the terms are logical or self-explanatory. When working through the book, refer back to this section if you come across an unfamiliar term.

1.5.3 Construction Lines

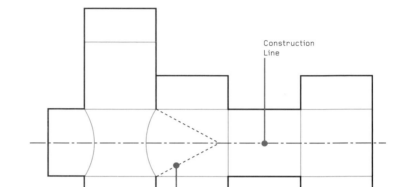

Construction Line

Radius

1.5.1 Box

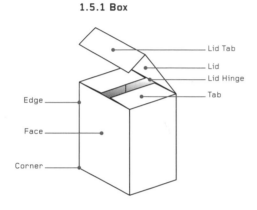

Lid Tab
Lid
Lid Hinge
Tab
Edge
Face
Corner

1.5.2 Valley and Mountain Folds

Valley Fold

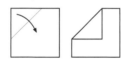

Mountain Fold

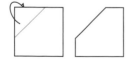

1.5.4 Net

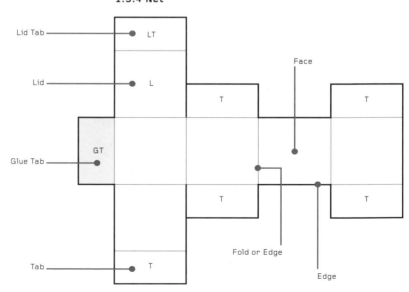

Lid Tab — LT
Lid — L
Glue Tab — GT
Tab — T
Face
T T
T T
Fold or Edge
Edge

1.5.5 Polygons

A polygon is a flat shape bounded by a closed path of straight sides. Any packaging form consists of a number of polygons, arranged in three dimensions. Some polygons – especially those with three or four sides – are subtly different one from another, and have different names. Knowing the names and understanding the differences will not only help you to understand the book better, but will also help you to design better.

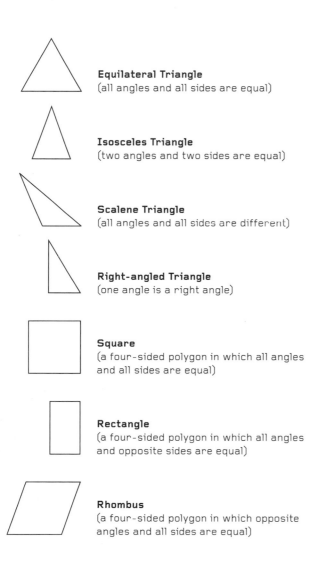

Equilateral Triangle
(all angles and all sides are equal)

Isosceles Triangle
(two angles and two sides are equal)

Scalene Triangle
(all angles and all sides are different)

Right-angled Triangle
(one angle is a right angle)

Square
(a four-sided polygon in which all angles and all sides are equal)

Rectangle
(a four-sided polygon in which all angles and opposite sides are equal)

Rhombus
(a four-sided polygon in which opposite angles and all sides are equal)

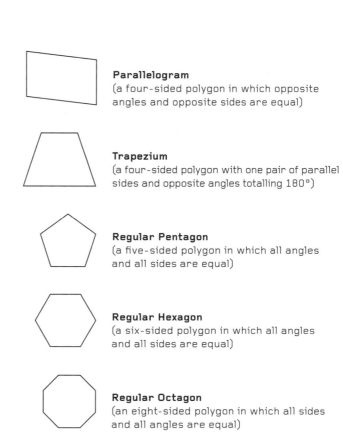

Parallelogram
(a four-sided polygon in which opposite angles and opposite sides are equal)

Trapezium
(a four-sided polygon with one pair of parallel sides and opposite angles totalling 180°)

Regular Pentagon
(a five-sided polygon in which all angles and all sides are equal)

Regular Hexagon
(a six-sided polygon in which all angles and all sides are equal)

Regular Octagon
(an eight-sided polygon in which all sides and all angles are equal)

02:

HOW TO DESIGN THE PERFECT NET

Introduction

This chapter is the core of the book. It describes in detail how to design a strong, one-piece, self-locking net to enclose any polyhedron (a three-dimensional figure with straight edges and flat faces).

The system it describes is precise and exacting and must be followed accurately, almost to the point of obsession – at least at first. Later, when you are familiar with it, you may take a short cut here, miss a step there, but at first it is necessary to learn it thoroughly.

Time spent on this chapter will be well rewarded. The longer you spend with it, the more you will understand when you come to design your own packaging – and the more innovative and practical this will be. Skip lightly over this chapter and your ability to design will be compromised. Sometimes, creativity comes from thinking freely without limitations, and sometimes it comes from learning something thoroughly and then applying it. Structural packaging is definitely in the latter category.

So please work slowly through this chapter; read it carefully and, if you have the time, make the examples. The chapters that follow use what it teaches, so understanding the principles of net design described in the following pages will enable you to understand how complex nets are constructed, and how you can use or adapt them.

Step 1:

By making drawings and rough 3-D models, decide the form of the package you want to make. This is the creative step!

This first step is the most important. If your design is poorly conceptualized, the most perfectly made net will not save it from criticism. It is crucial to spend as much time as possible drawing, making quick 3-D models and discussing ideas and results with colleagues, so that you are confident that what you have designed in rough is ready to be taken through the sequence of technical net construction steps that follow.

If you are looking for ideas, use the book for inspiration. The latter half in particular contains many interesting packaging forms which are probably not exactly right for your needs, but which can be adapted or combined to create something original, using the principles of net construction explained in this chapter.

You should not begin Step 2 until you are confident that your roughly made package (or box, tray, bowl, display stand, sculpture or whatever) is absolutely the right design.

Remember: this book does not tell you **what** to design, but **how to make** what you have designed.

Step 2:

Using one sheet of card for each face, construct the package as a solid brick. Hold the faces together with masking tape. Give no thought to the net, the lid or the tabs.

Make each face carefully from a sheet of card. This can be done either by hand using geometric construction equipment, or by using a CAD or graphics application and making printouts of the faces. If you are unsure of the dimensions of your packaging, this step will fix them, though they can always be changed later.

Use masking tape to fix all the faces together strongly, edge to edge. (Masking tape is a low-tack beige-coloured paper tape, widely available from office suppliers, home improvement stores and art/craft retailers.) Avoid using a plastic tape, as you will need to write on the tape in Step 3. The result should be a well-made, sturdy dummy of your package held together with tape.

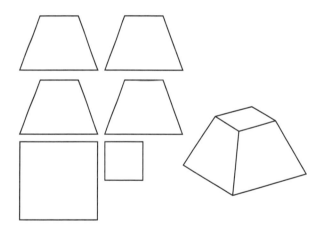

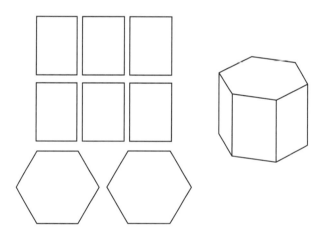

Example 1
Four trapeziums and two squares create a truncated pyramid. The length across the top of each trapezium is the same as the side length of the small square. The length across the bottom of each trapezium is the same as the side length of the big square. The height and slope of the trapeziums are unimportant, but if you are copying this design as a learning exercise, make the trapeziums look somewhat like those shown here.

Example 2
Six rectangles and two hexagons create a hexagonal prism. The height of the rectangles is unimportant, but their shorter sides are the same length as the sides of the hexagons.

Step 3:

Write pairs of identical numbers across each edge.

These pairs of numbers locate the position of each face in relation to all the other faces, so that if the pieces were separated, the package could be assembled again like a 3-D jigsaw. More importantly, the numbers also show which edge on which face touches which other edge on which other face. Knowing which edges touch means the tabs can later be added in the correct places.

For clarity, write the numbers large and in the approximate centre of an edge. There is no logic to the numbering system; the edges can be numbered in any sequence, no matter how random.

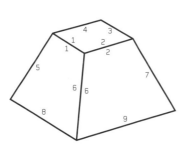

Example 1

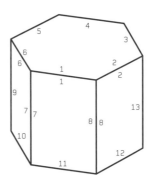

Example 2

Step 4:

If the package has a lid, cut it loose.

Depending on what you are making, your 'package' may not be a package
at all, but a 3-D form with another function. If so, you may not need a lid
and can skip this step. But if your 3-D form is indeed a package, it probably
will have a lid. The shape and position of the lid would have been decided
in Step 1.

With a sharp knife, cut carefully through the masking tape to release the lid,
leaving it joined to the remainder of the package along one edge. Cut through
the tape rather than removing it, as removing it may pull off the numbers you
added in Step 3.

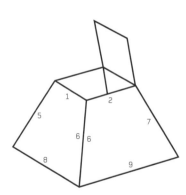

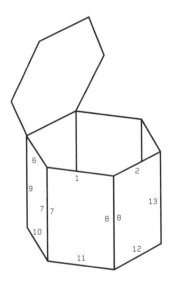

Example 1
The most sensible face on the package
for a lid is the small square, though the
big square would give easier access to
the interior.

Example 2
A hexagon is an obvious face for a lid,
though it would be more interesting to
create a lid from one of the rectangles.

Step 5:

Using masking tape, affix a tab securely to the lid edge that is opposite the hinge. If no edge is opposite, choose another edge instead. The tab should have corners of 90°.

This first tab is called the 'lid tab' and is the most important tab on the net because it determines the positions of all the other tabs.

The temptation is to make it too skinny, but instead, be generous and make it quite deep. It is easier to trim it narrower later than to remake it deeper. Fix it securely to the lid with masking tape, front and back.

The tab may need corners with angles of less than 90° if it is to be fitted into a tapering face. The 'Troubleshooting' section on page 32 will help you. On no account make the corners of the tab bigger than 90°; if you do, it will not slide in and out of the package easily.

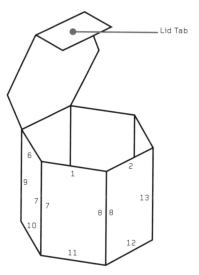

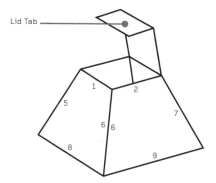

Example 1
The lid tab is placed in a conventional position on the lid.

Example 2
The lid tab is placed on a hexagonal lid so, unusually, there are two empty edges to the lid left and right of the tab on the way back to the lid hinge.

Step 6:

Cut loose as many of the shortest edges as you can.

Pick up your package and examine it carefully. Make a mental note of which edges are the shortest. There may be just one or two of them, or perhaps quite a few of equal length.

Then cut through as many of those shortest edges as you can without releasing a face completely from the others so that it falls off. It's not important which edges you cut or leave uncut, but it helps to try to work symmetrically, doing the same cutting top and bottom, or left and right, around the form.

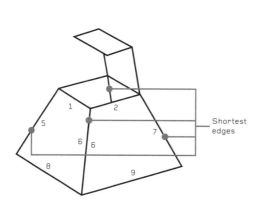

Example 1
The shortest edges are all the sloping edges of the trapeziums.

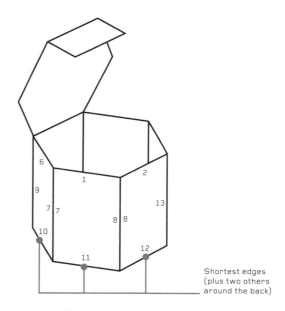

Example 2
Cut through five of the six short edges around the bottom of the package. The sixth edge is left uncut. Try to leave this uncut edge directly under the lid hinge, so that in Step 7 the hexagons are in line, one beneath the other.

Step 7:

Now, cut open the remainder of the package until it can be laid out flat. Begin by cutting loose the shortest edges that remain uncut, then cut loose progressively longer and longer edges.

This is a critical step because, for the first time, your design has transformed from a 3-D form into a 2-D net. It may be that you make a mistake or two in the cutting, by cutting long edges when you should have cut shorter ones. If so, reassemble the package to create a 3-D form, and apply masking tape to join together edges that were mistakenly separated. Then cut other edges loose. If during this process you become confused as to which edge touches which other edge, the number pairings will keep the faces and edges correctly aligned.

If your package has a large number of faces, there will be a very large number of ways in which it can be cut open to become flat. These options will be limited by cutting the shortest edges first (Step 6), then by cutting progressively longer edges (this step), but even so, there will still be many options. In the end, there may be no single 'perfect' net, but a few, or even many, nets each of which is as good as the other.

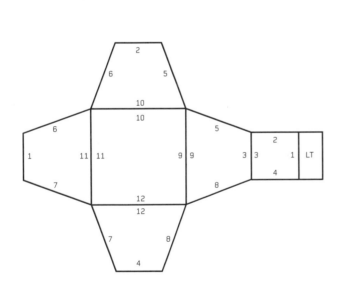

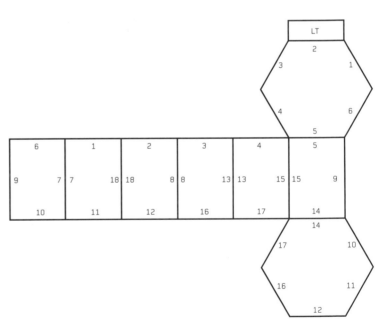

Example 1
This is the net for a truncated trapezium. Here there are no variations that would look significantly different.

Example 2
The two hexagons can be joined to the line of six rectangles in many different places. These positions are all as good as each other, but with the net shown here, edge 9 touches the other edge 9 where the lid joins the rectangles. This means that if the box is printed on (which is perhaps likely), the printed image has the benefit of being continuous around the 'front' of the box.

Step 8:

Adjacent to the number, write a T (for tab) near the edge of the lid next to the lid tab.

This simple step begins the identification of which edges should be tabbed and which should remain untabbed. Write the T clearly, next to the number that has already been written.

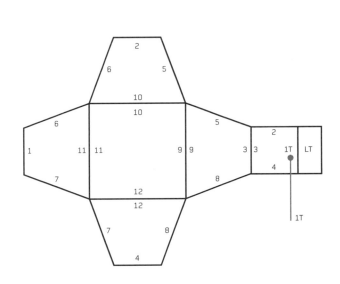

Example 1

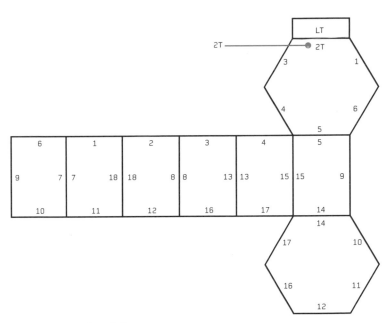

Example 2

Step 9:

By marking each edge around the perimeter with a T or an X, the net can be prepared for tabbing. The first T has just been written, so write an X on the next section of the perimeter, then T again, then X again ... Continue with the T-X-T-X pattern until all the perimeter has been lettered. Write the letters next to the existing numbers.

On the edge adjacent to the lid tab (which has previously been marked with a T), write an X next to the number. On the next edge write a second T, then on the next edge write a second X. Continue around the perimeter marking every edge with alternate T and X symbols, to create a T-X-T-X-T-X-T-X ... etc. pattern around the perimeter. Write the letters alongside the numbers, such as 4T or 7X. If you do it correctly, the last edge you mark will have an X symbol, adjacent to the T symbol on the lid tab, made in Step 8. In this way, the pattern has no beginning and no end. Every net, no matter how eccentric or complex, will have an even number of edges, so the T-X pattern will always work.

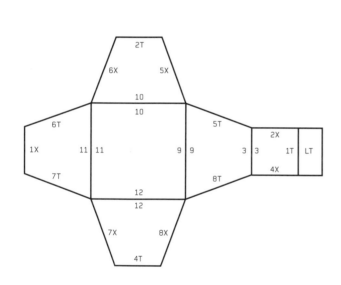

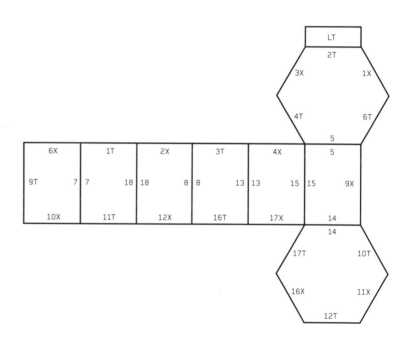

Examples 1 & 2
The T-X tabbing patterns are shown complete. If in Steps 4 and 6 you damaged some of the masking tape and compromised the legibility of any numbers, write them again clearly so that everything can be read with ease.

Step 10:

There are two aspects to tabbing a net. The first is the correct placement of the tabs, which is described in this step.

The edge to which the lid tab was fixed was marked with a T. The remaining tabs will affix to all the edges marked with a T. In this way, the tabs are placed on alternate edges around the perimeter.

This is the core of the system for correctly tabbing any volumetric form. Two or more tabs are never placed adjacent to each other, nor are there ever two edges or more between tabs. Providing the net has been correctly made in the preceding steps, the tabs will automatically be placed in the correct positions to lock together in the strongest possible way, when any volumetric form is folded up from one piece of card.

The positioning of the lid tab dictates the position of every other tab. However, if there is no lid and therefore no lid tab, the tabs could equally well be placed on all the X edges as on all the T edges.

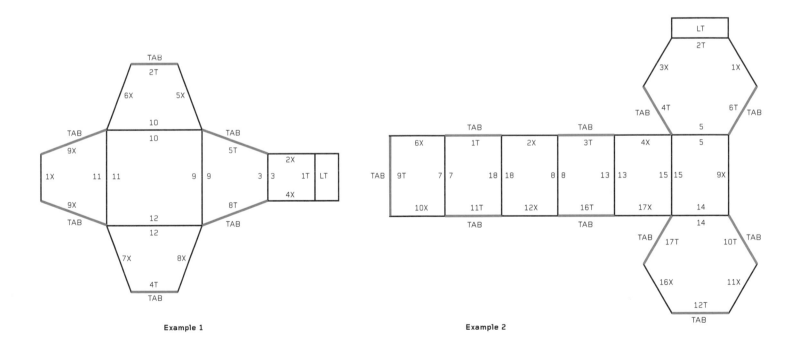

Example 1   Example 2

2.   HOW TO
      DESIGN THE
      PERFECT NET
2.11 Step 11.1

Step 11:

The second aspect to correctly tabbing a net is defining the shape of each tab. Consider each number pairing made in Step 3. It will be seen that around the perimeter of the net each pair has one T number and one X number, such as 4T and 4X. In step 10 a tab was placed on each T edge. When folded up, the tab on edge 4T will slide inside edge 4X. Thus, the tab on edge 4T must be the same shape as the face beyond edge 4X. This principle applies to all the number pairs on the perimeter.

If the first aspect (Step 10: the placement of the tabs) is simple to understand, this second aspect (the shape of the tabs) is perhaps more subtle. The shape of each tab must be decided individually. There is no quick way to do this — every tab must be designed carefully and accurately, one at a time.

Here is the method, step-by-step.

11.1

The tab that will join to the T edge must be the same shape as the face beyond the corresponding X edge. If necessary, measure the angles of the face at the ends of the X edge, as this will determine the shape of the face and, later, of the tab.

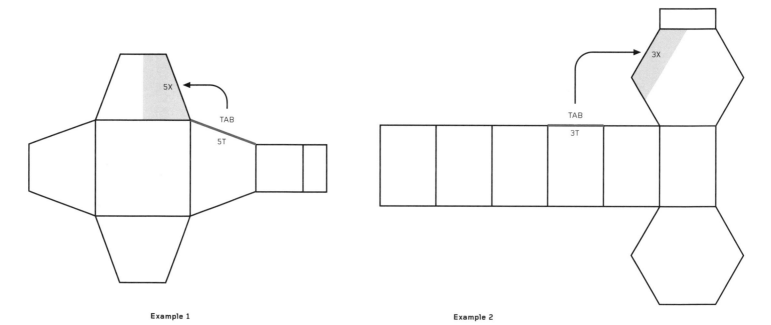

Example 1   Example 2

11.2
This is not a step, but a graphic to help you visualize how the shape of the face beyond the X edge joins to the T edge to become the tab. In both examples, the face is seen floating away from its position beyond the X edge, towards its location on the corresponding T edge.

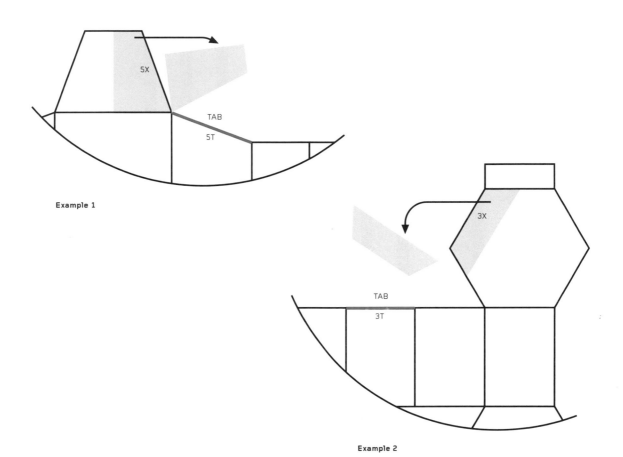

Example 1

Example 2

11.3
This is the face from beyond the X edge, now copied on to the corresponding
T edge as the tab.

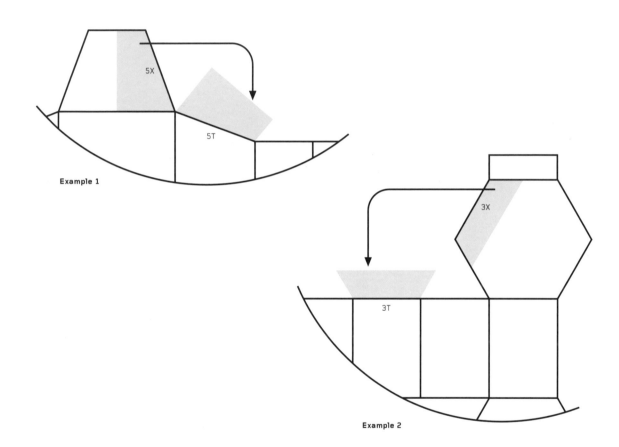

Example 1

Example 2

11.4
Cut out the shape of the tab from card and affix it securely to the T edge with masking tape, front and back. Notice that the tab is cut quite deep, not thinly. Check that when the 2-D net is folded up to make a 3-D package, the tab will fit exactly into the face beyond the X edge. If it doesn't fit, simply remove the tab and remake it.

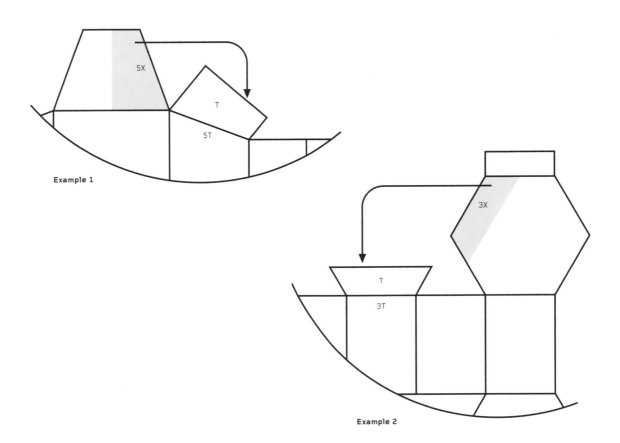

Example 1

Example 2

11.5

Repeat the same procedure with all the remaining tabs. If the box is complex
with many different lengths and angles, this may be a lengthy procedure,
but it is vital to do it methodically and accurately. Take your time.

This step is the core of the net construction system. Followed with 100 per
cent accuracy, it creates a net of remarkable strength. Done with even
99 per cent accuracy, the net will be weakened. A net is either absolutely
correct and perfect, or incorrect and in need of correction. In design, the
concept of perfection is almost unknown – how can a magazine layout, or
a colour, or a choice of fabric be described as perfect? – but in package
design perfection is achievable and necessary.

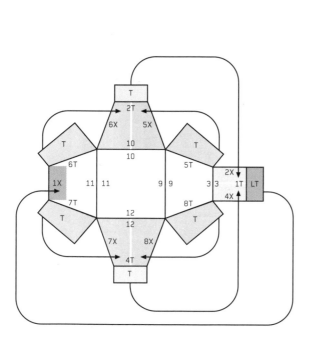

Example 1

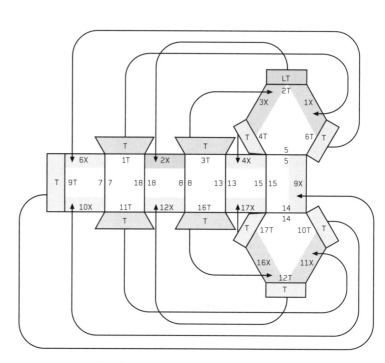

Example 2

11.6
This is the completed net. When all the separate tabs have been checked for accuracy and you are confident that you have created an accurate collage of your design for a package, it can be remade from one sheet of card.

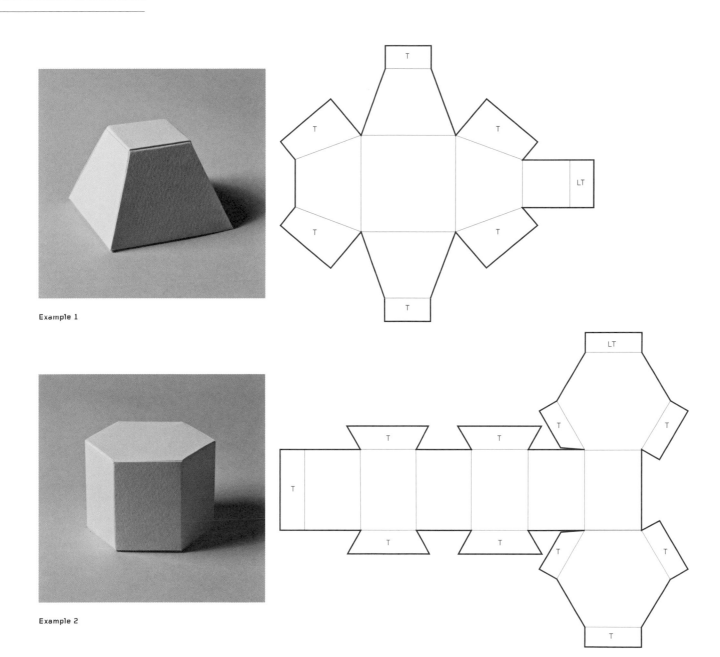

Example 1

Example 2

2.   HOW TO
       DESIGN THE
       PERFECT NET

2.12 Troubleshooting

2.12 Troubleshooting

The method described in the previous pages is very precise and must be followed accurately. Any deviation will weaken the final net. While following the steps, it is possible to make small mistakes which compromise the strength and integrity of what you are designing. Here, then, are answers to some common problems.

Q: At Step 7, I cut my package flat, but made a poor net. What should I do to improve it?

A: In truth, at Step 7 many nets will need a little improving and this must be done before progressing on to Step 8. The way to improve a net is to redesign the arrangement of the faces.

To illustrate the method of redesigning the net, here is an extreme example of how a poor net can be made good. The long, cuboid box was made correctly following the method given in Steps 1–6, then cut open incorrectly in Step 7 to make a flat net.

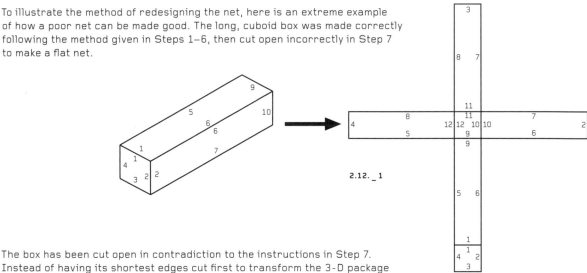

2.12._1

The box has been cut open in contradiction to the instructions in Step 7. Instead of having its shortest edges cut first to transform the 3-D package into a 2-D net, the longest edges were cut first. This has created a net that is fragile and which occupies a very large rectangular area of card, much of which will be wasted. The package made from this net will therefore be weak and expensive.

The following sequence shows how faces can be cut from the net and rejoined at better places. The intention is to join together as many of the longest edges as possible, to create a net with as many of the shortest edges as possible around the perimeter. To do so will create the strongest possible package from the minimum rectangular area of card, thus maximizing the strength and minimizing the cost of the design.

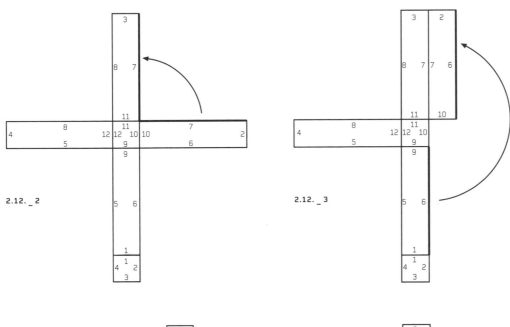

2.12. _ 2

2.12. _ 3

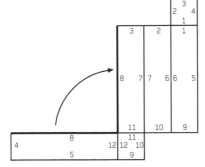

2.12. _ 4

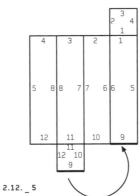

2.12. _ 5

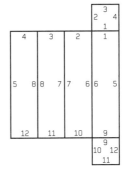

2.12. _ 6
The net is now maximized for strength and compactness and can be tabbed according to Steps 8–11.

Q: I have followed the steps correctly, but my package will not lock. What can I do?

A: There are several possible reasons for this. Any one of them, or perhaps a combination of two or more, may be the cause.

1. Is it made well?
A poorly made package will not lock. Check that the faces and the tabs are precisely made and that everything has been cut or folded with accuracy and attention to detail. Check, too, that the card is not too thin or too thick (250gsm is the weight used in this book, though you may eventually wish to use something heavier) and that the folds are not too floppy and without strength.

2. Are the tabs deep enough?
A package will hold together without glue because the tabs fit snugly inside. The side edges of each tab rub against the inside of a folded edge, so the longer the side edges are, the more grip the tabs will have and the more strongly the box will hold together.

Here are two good nets for a simple cube. The net with the deep tabs will be considerably stronger than the net with narrow tabs. It is conceivable that the latter will not hold together.

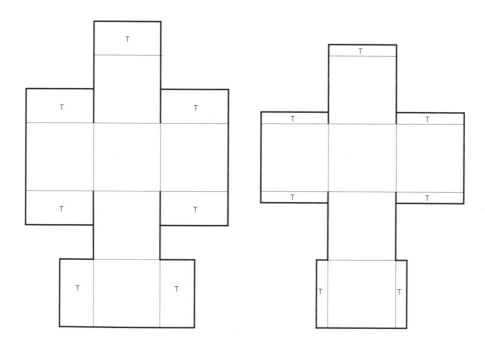

3. Tapering tabs

In the examples below, the 90° tabs will lock the cube very strongly. However, the tapering 60° tabs will not lock the tetrahedron, although they are correctly positioned and correctly shaped.

There are three remedies to this problem:

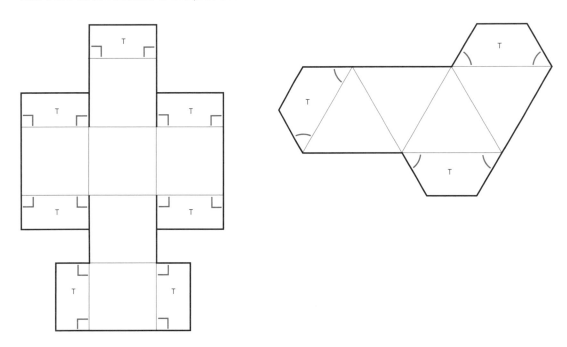

Remedy A

Glue the tabs! While totally possible, this is not the recommended method. Remedies B and C offer better solutions.

Remedy B

Add flanges to the edges of the tapering tabs. A flange is an extra piece of card joined to the side of a tab that has a corner of less than 90°. Adding flanges will help to hook the tab around others inside the package, and so lock the tetrahedron securely together.

Here is an example of a simple three-sided pyramid, known in geometry as a tetrahedron. Following the steps above, a simple net is made with three tapering tabs of 60°. These tapering tabs will not lock the tetrahedron together, so flanges need to be added. The size of the flanges will depend on the size of the package and the weight of the card. It may be enough to extend the tabs only by an extra 30°, so that they become 90°, or perhaps by an extra 60°, so that they become 120°. It is better to make the flanged tabs 120° rather than 90°, and to trim off any excess. Too big is better than too small.

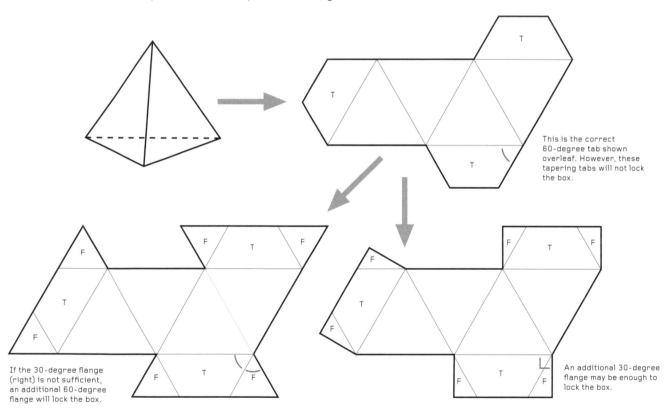

This is the correct 60-degree tab shown overleaf. However, these tapering tabs will not lock the box.

If the 30-degree flange (right) is not sufficient, an additional 60-degree flange will lock the box.

An additional 30-degree flange may be enough to lock the box.

Remedy C

Add a Click Lock (see page 70) to each 60° tab. This is a simple, secure lid lock, found on many mass-produced cuboid boxes. It will also securely lock tabs that taper.

2.12   Troubleshooting

Q: I have no space on my net for some of the tabs. Where can I put them?

A: With complex nets it is quite common for sections of the perimeter to become so crowded in a few places that it is impossible to create tabs that are wide enough and deep enough to be effective. In these instances, the solution is to follow the answer to the first 'Troubleshooting' question (see page 32) and move the faces around on the flat net. Remember to keep the longest edges connected and to keep any lid tab in the correct place. Never omit a tab or leave it unchanged knowing it is incorrect and in need of improvement.

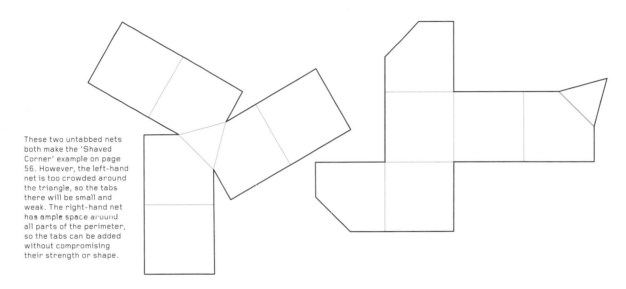

These two untabbed nets both make the 'Shaved Corner' example on page 56. However, the left-hand net is too crowded around the triangle, so the tabs there will be small and weak. The right-hand net has ample space around all parts of the perimeter, so the tabs can be added without compromising their strength or shape.

Q: Almost all my edges are the same length. In Step 7, which should I cut first and why?

A: If you have a wide choice of edges to cut, consider these criteria to help you narrow your options.

1. Cut open the net to make it occupy as compact an area as possible. This will consume less card and make the package less expensive to manufacture.

2. Cut open the net to leave ample space for all the tabs (this is the answer given above on this page). This will make the package stronger.

3. Cut open the net in a symmetrical way, rather than in some random configuration. This usually helps to make the net stronger and more compact.

4. Keep certain key edges connected so that if the surface is printed on, this can be done without interruption, across the folds connecting one face to another. This will improve the appearance of the surface graphics.

03:

SQUARE-CORNERED BOXES

Introduction

Square-cornered boxes are a ubiquitous staple of modern life, so everyday that we barely stop to look at them and never think to praise them.

They are, however, a design classic: they are easily morphed in three dimensions; the nets have little wastage; they can be machine-assembled; they can be stored flat and erected instantly; they tessellate fully in three dimensions for economic transportation, storage and display; they have large, flat areas for printing; and when their useful life has come to an end, they can be recycled.

This chapter presents the ten basic ways in which a square-cornered box can be designed, depending on the dimensions of the sides and the placement and orientation of the lid.

The uniformity of the angles at the corners of a square-cornered box – they are all 90° – means that in many ways, these are the simplest boxes to construct. That is why they appear first in a series of chapters that present nets of increasing complexity and creativity.

Throughout, it is assumed you have read Chapter 2 and understood the system of net design and tabbing that it presents. The lids shown here all use simple square tabs without rounded corners. However, you may wish to use rounded corners or instead, a Click Lock (see page 70) or Tongue Lock (see page 72) to close a lid more securely. All the nets show a glue tab or glue line (see page 68). The boxes will lock very well without glue, but will 'explode' open once the lid is loosened if they are assembled unglued.

3.1 Which Net?

The six squares and/or rectangles which join together to make a square-cornered box can be configured in 11 different ways, excluding mirror images and rotations. The number doubles to 22 once the box is tabbed, depending on whether the tabs are placed on the T or X edges (see page 24).

Which net is the best? There may be no single answer, but many of the 11 (or 22) possible nets can be quickly dismissed when the following criteria are considered:

1. Which face is the lid and which edge is the lid tab?
2. Do certain faces need to be connected to provide a continuation of the printed image?
3. Is the net as compact and as strong as possible?

Below are the 11 possible nets from which a cube or cuboid box can be created, excluding tabs. Some of them may already be familiar to you, but others are rarely used and rarely seen. Notice how each net has at least two squares with three raw edges and only one folded edge (some nets have three or four such squares). One of these squares must become the lid.

3.2 The Basic Cube Box (A x A x A)

A cube has six identical square faces. In its three dimensions of height, breadth and depth the sides are the same length (A). Thus, a cube can be described as an A x A x A form.

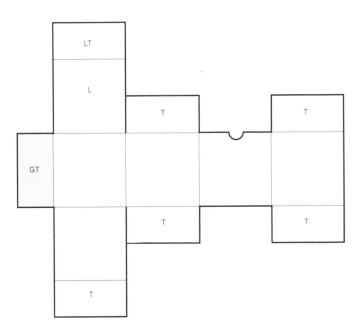

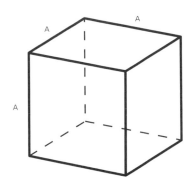

Here is one possible way to make a net for a cube, using the system described in Chapter 2.

3.3 Square Cuboid Boxes (A x A x B)

A square cuboid has a square cross-section (A x A), but the third dimension (B) can be any length greater or smaller than A. There are three basic A x A x B cuboid boxes, depending on the placement and orientation of the lid, though because B is of no fixed length, there are many variations on these.

3.3.1 A x A Lid

An A x A x B box of this configuration is strong and stable, though if B lengthens significantly, the optional glue line becomes essential if the box is to retain its strength. B may also be shorter than A, to create a squat box with a square lid.

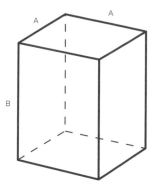

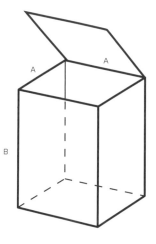

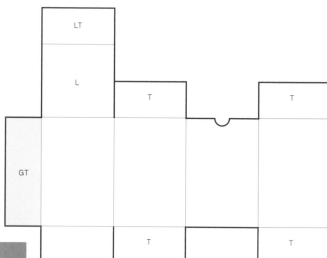

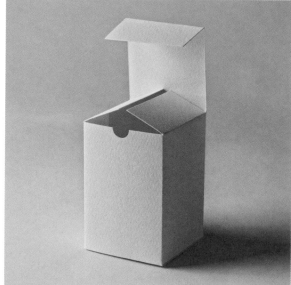

3.3.2 A x B Lid, with a B Hinge

The long front edge of this lid means that a Click Lock (see page 70) is advisable to hold the lid securely closed. If the raw edge in front of the lid tab bows too much, a Tongue Lock (see page 72) would be a preferable alternative.

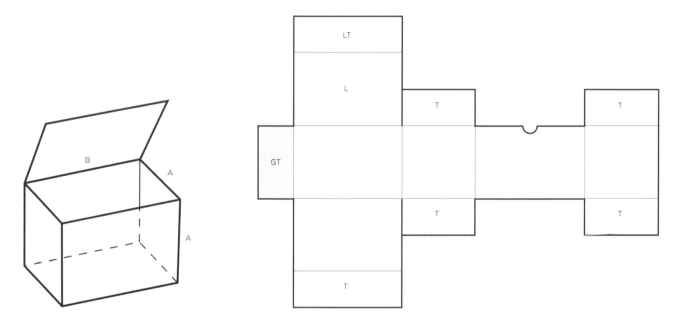

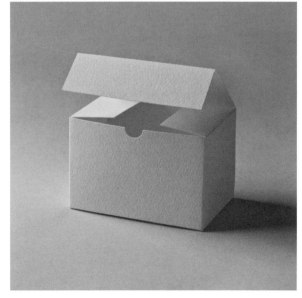

3.3.3 A x B Lid, with an A Hinge

Of the three A x A x B configurations, this is the least satisfactory. The long lid opening is vulnerable to damage. Also, to make the net more compact the base of the box has been turned sideways to connect its longer edge to the main body of net; this may mean that if a glue tab is made, the tabs around the base of the box will become difficult to interlock. The solution would be to configure the net so that the base is a mirror image of the lid.

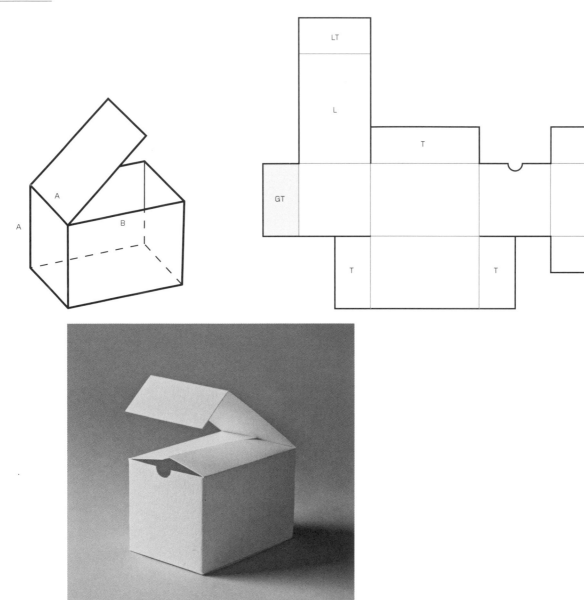

3.4 Rectangular Cuboid Boxes (A x B x C)

A rectangular cuboid box is a box in which all three dimensions (A, B and C) have different lengths, so all the faces are rectangles. Opposite faces are congruent (equal in shape and size). There are six possible lids, depending upon whether the lid hinge is on the long or short side of one of the three different rectangles. With all the three lengths variable against each other, the number of possible nets becomes immense, though they will all be a variation of one of the six nets shown on the following pages.

The nets may appear somewhat repetitious, but close study will reveal them to be subtly different. They are so fundamental to a study of package design that they are well worth documenting in full. Some of the lids will benefit from an additional Click Lock (see page 70) or Tongue Lock (see page 72), depending on the size of the box and the material used to make it.

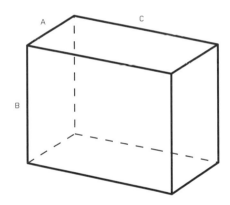

3.4.1 A x B Lid, with an A Hinge

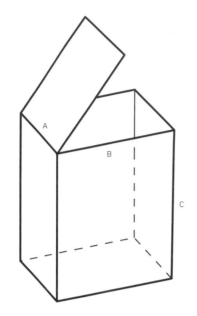

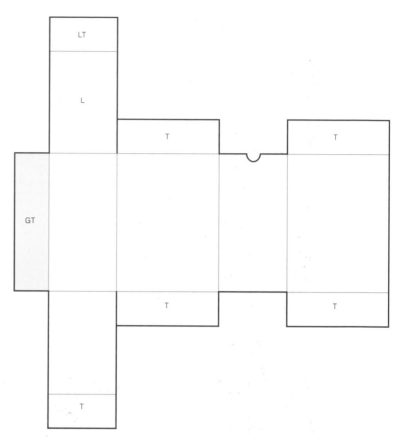

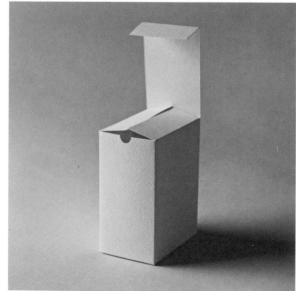

3.4.2 A x B Lid, with a B Hinge

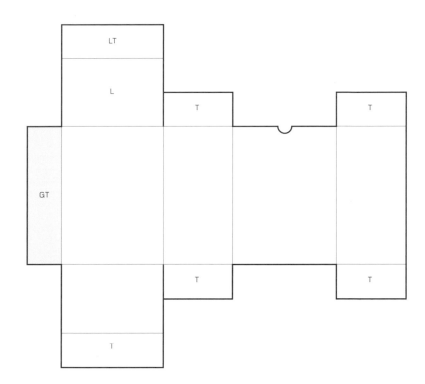

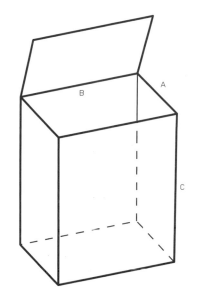

3.4.3 A x C Lid, with an A Hinge

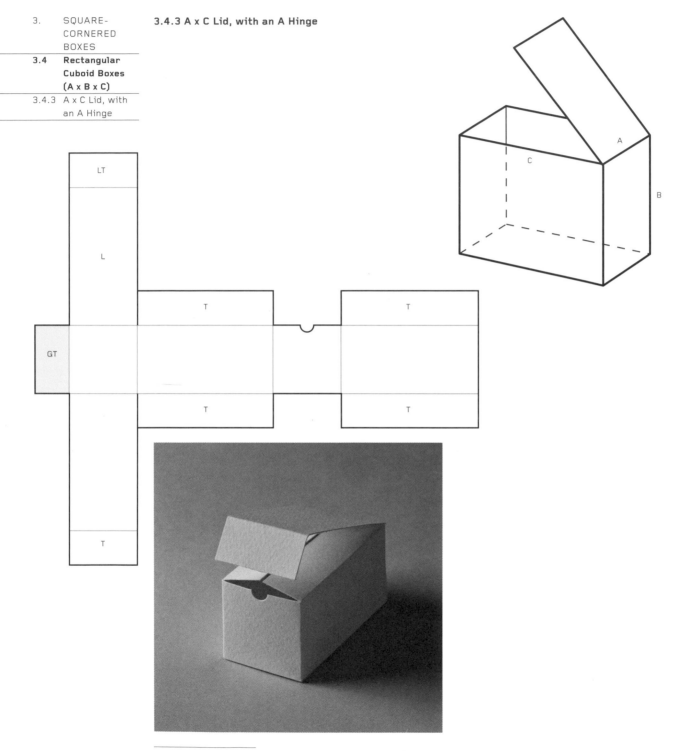

3.4.4 A x C Lid, with a C Hinge

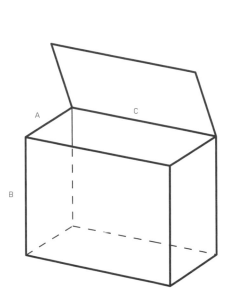

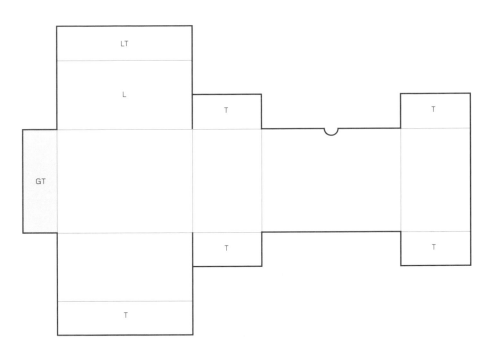

3.4.5 B x C Lid, with a B Hinge

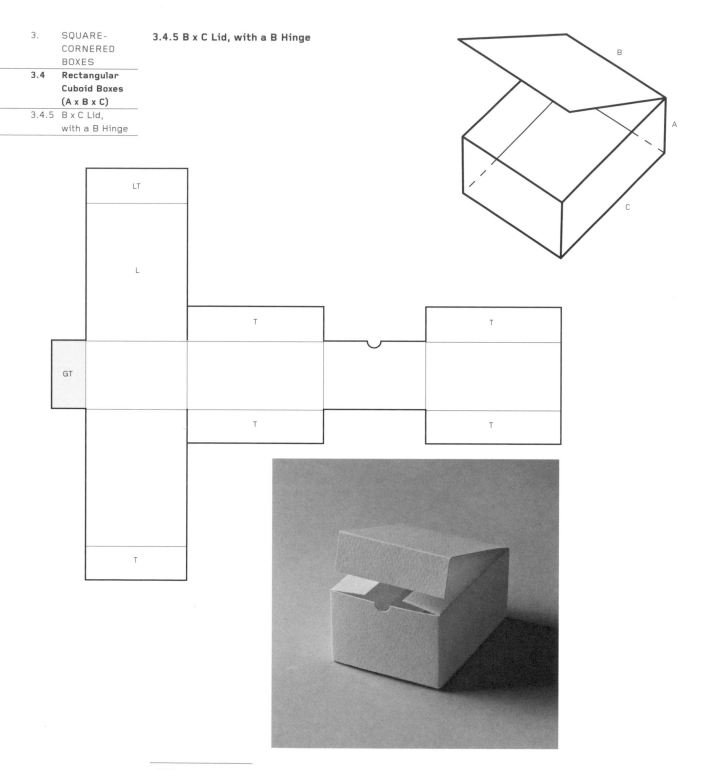

3.4.6 B x C Lid, with a C Hinge

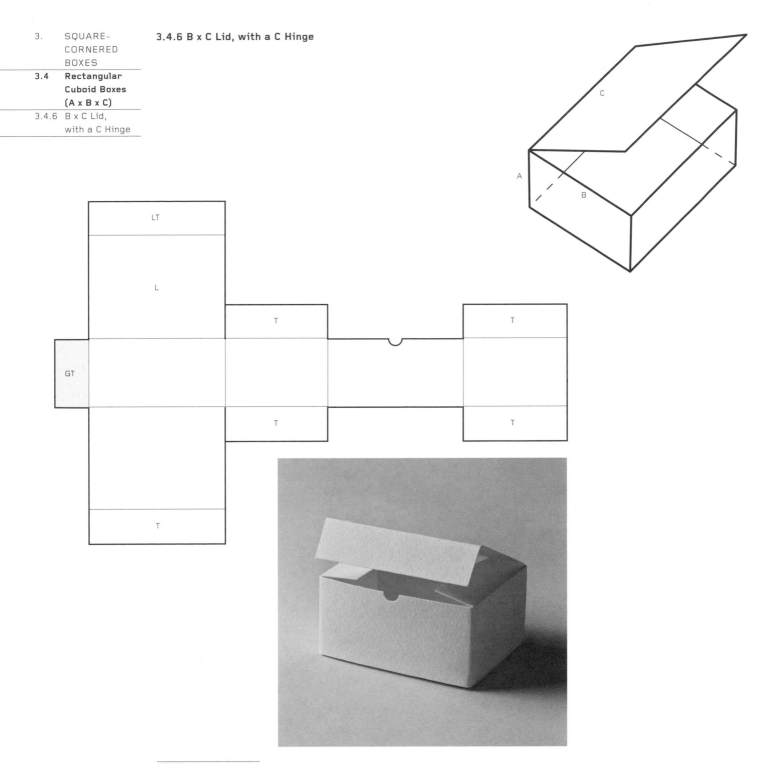

04:

DEFORMING A CUBE

Introduction

A cube can be deformed in many simple ways to create a large number of forms suitable for packaging. Like almost all examples of structural packaging, a cube consists of three elements:

- faces
- edges
- corners

Imagine that a cube is made of solid wood. A little of the wood may be shaved away, or a lot, symmetrically or asymmetrically, at one face, edge or corner, or two or more, to create a large array of new forms. Each resulting form can be made as a net following the method explained in Chapter 2.

Further deformations can be made by imagining the cube is solid rubber, which is stretched, compressed or twisted along axes connecting opposite faces, opposite edges or opposite corners.

Finally, certain straight edges can, in some instances, be replaced by curved edges, which in turn may induce the previously flat faces to curve.

This chapter presents some of the most basic ways in which a cube can be deformed using the strategies mentioned above. When these principles are understood, it becomes surprisingly easy to create new packaging forms, even if your knowledge of three-dimensional geometry and polyhedra is minimal.

Throughout, the chapter assumes you have read and understood Chapter 2, so you understand how the nets have been constructed – particularly the configuration of the faces relative to each other and the shape and placement of the tabs. In some instances, the designs may benefit from a Click Lock (see page 70), a Tongue Lock (see page 72) or a Glue Tab (see page 68).

None of the forms have designated lids. Each form could have a lid placed on one of several faces, in a variety of orientations. To present them all would simply take too much space in the book. In any case, following the method of net construction explained in Chapter 2, it is a simple matter to redesign a net so that the lid appears on the face of your choice.

At least one basic deformation is missing: the truncated pyramid from Chapter 2, which merits a place here. If you are making the set, please refer back to that chapter.

4.1 Shaving a Face

A face of the cube is shaved away at an angle, creating a sloping top. The angle of the slope may be shallower or deeper than the angle shown here. It could also tilt forwards or backwards to create a more complex slope.

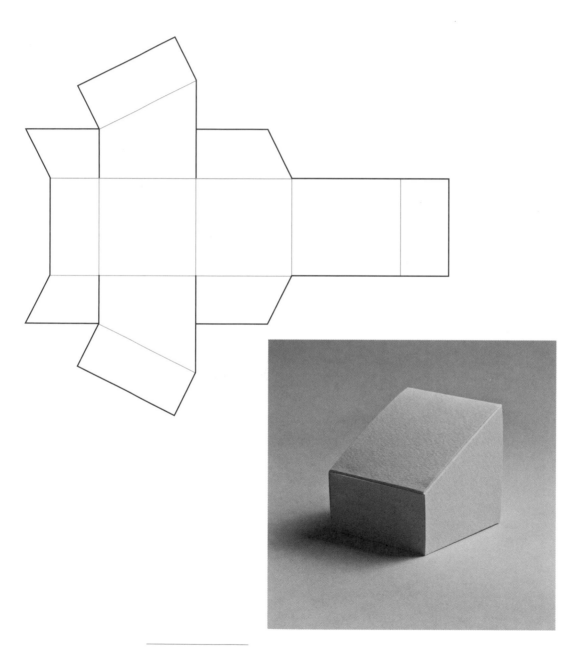

4.2 Shaving an Edge

An edge of the cube is shaved away to create an extra face. The amount shaved away could be less or more than the amount shown here and could tilt forwards or backwards.

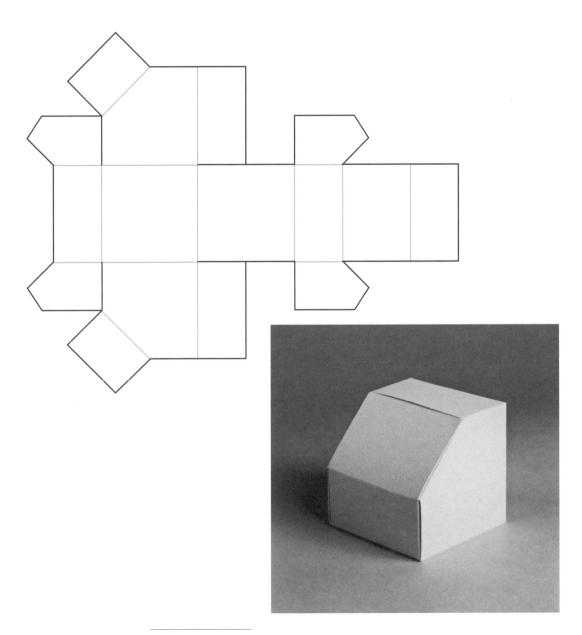

4.3 Shaving a Corner

A corner of the cube is shaved away to create an extra face. Perhaps surprisingly, the new face is an equilateral triangle (all the angles are 60°). As with 4.1 and 4.2, the amount shaved away could be more or less than the amount shown here. The triangle need not be equilateral. For a special sculptural effect, try standing the cube on the triangular face!

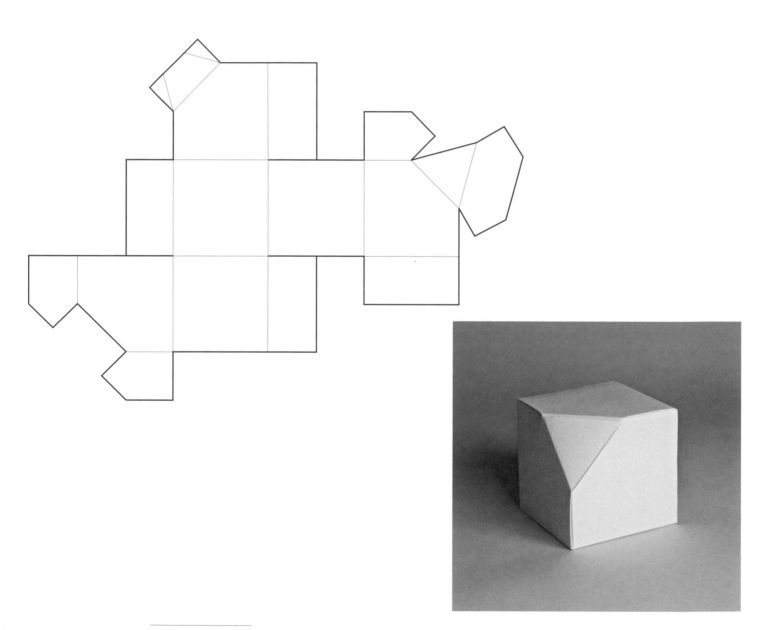

4.4 Stretching Edge-to-edge

A cube made from solid rubber could be stretched by holding a pair of opposite edges and pulling. The result would be the leaning box shown here.

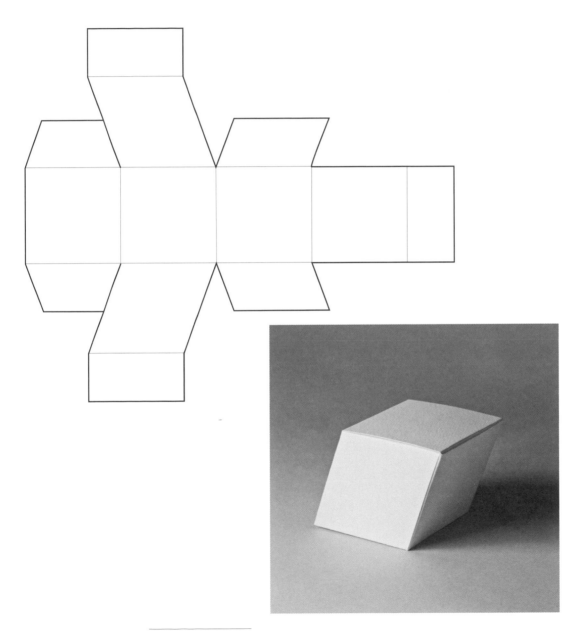

4.5 Stretching Corner-to-corner

In a similar way to 4.4, a rubber cube could be stretched by holding a pair of opposite corners and pulling. The result is the simple diamond shown here. This is a very enigmatic shape that apparently defies the laws of perspective.

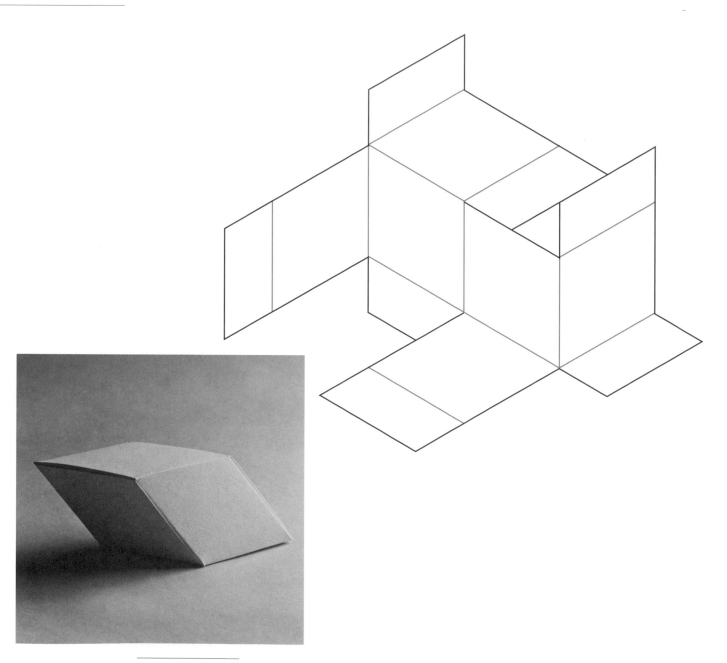

4.6 Twisting Opposite Faces

Our hypothetical rubber cube can be further contorted by being given a small twist, so that the top and bottom faces are twisted in opposite directions relative to each other. The resulting form is subtle and surprising. The angle of the skew is 80°, just 10° off the 90° perpendicular. Angles below 70° rarely work well.

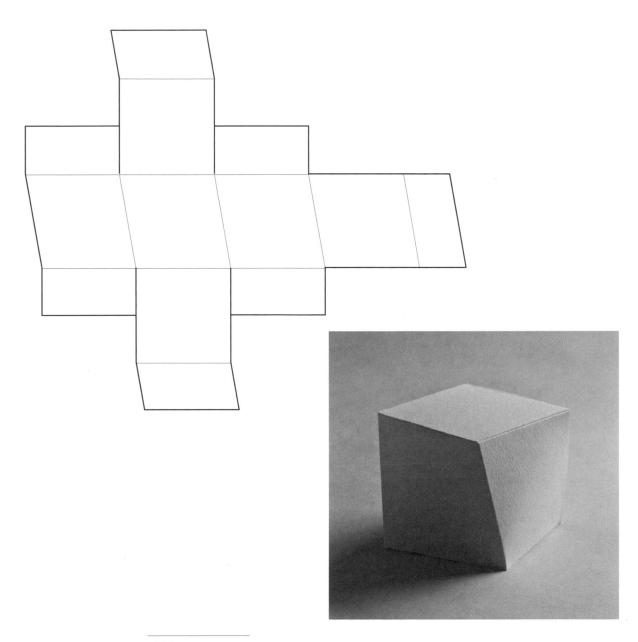

4.7 Twisting: Faceted Version

Example 4.6 can be twisted more and more until the corners of the top square lie directly over the midpoints of the edges of the bottom square. The eight corners of the two squares can then be connected with creases to create eight isosceles triangles around the sides. Note the flanged tabs at the top and bottom of the net, discussed in 'Troubleshooting' (see page 36).

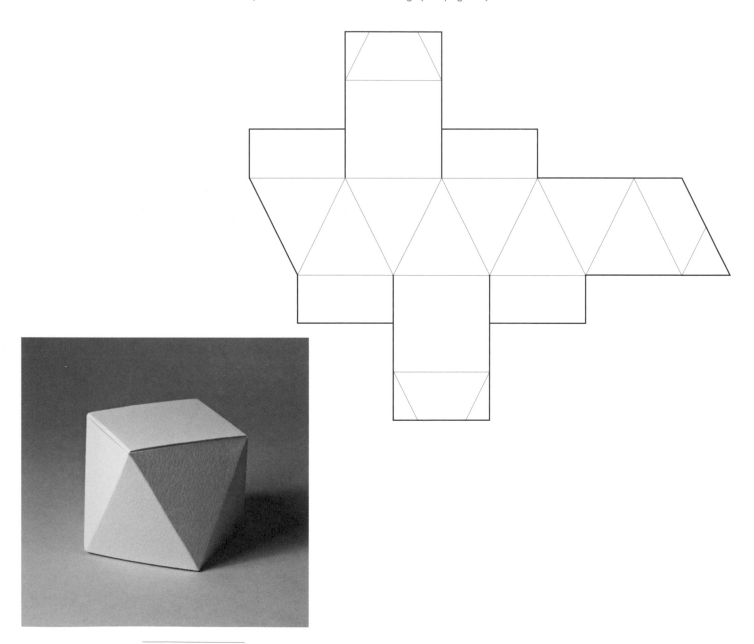

4.8 Compressing Face-to-face

Instead of being stretched or twisted, our rubber cube may be compressed.
The simplest compression is face-to-face. However, the resulting form
would not look like the image shown here – a stylization of a complex, curving,
jelly-like form. Although apparently compressed, the surface area of the card
is the same as that of a cube.

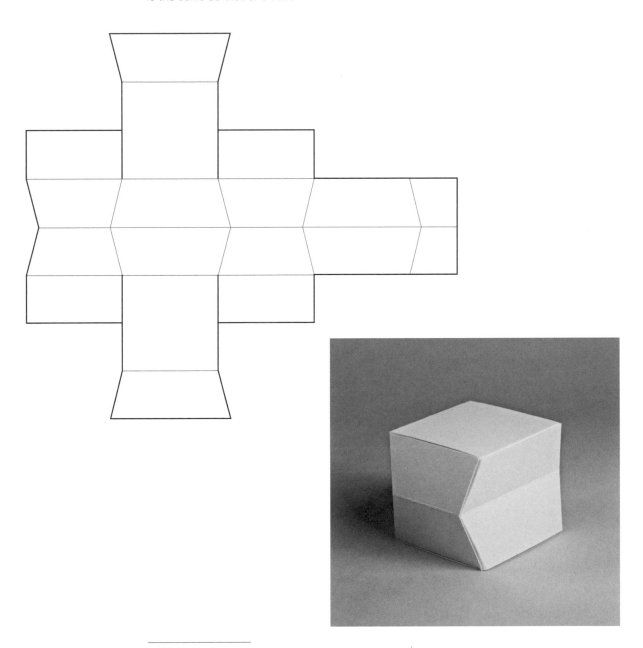

4.9 Double Curves

Almost any straight folded edge on any net can be substituted with a double curve to transform very hard-edged forms into ones that are softer and more decorative. These double curves need to be made very precisely, so here is the method in detail.

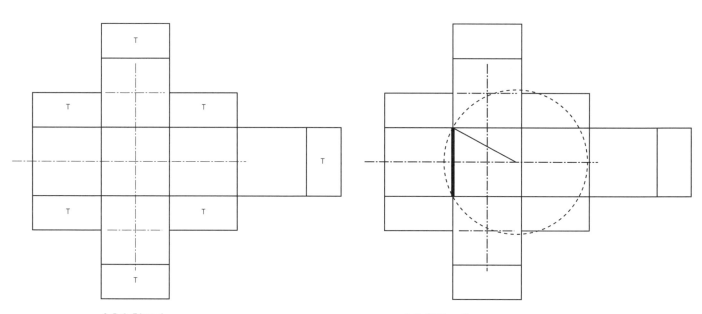

4.9.1 Step 1

Draw a standard net for a cube, including tabs. Add two long and two short construction lines as shown, across the centres of the squares.

4.9.2 Step 2

The thick line shows a straight folded edge on to which a double curve will be placed. Measure the length of the edge, then add 5 per cent. This will be the radius of the curve. Draw an arc of a circle of this radius, which passes exactly through both ends of the line. The centre of the circle is on a construction line to the right of the chosen line.

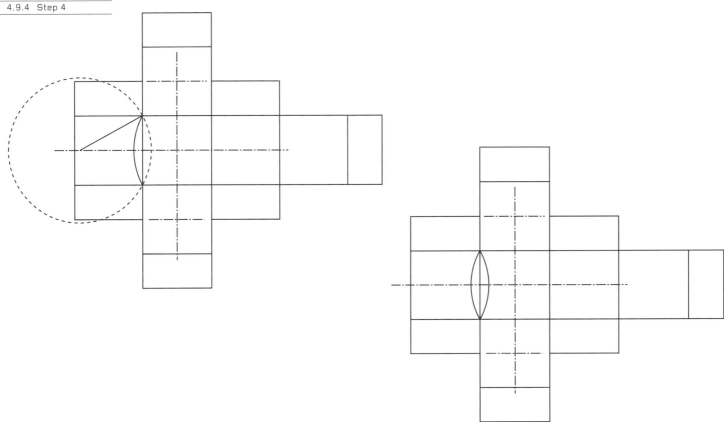

4.9.3 Step 3

The solid curved line was made in the previous step.
Now repeat Step 2, but this time to the left of the
chosen line.

4.9.4 Step 4

The double curve is now complete.

4.9.5 Step 5

Three more double curves have been added using the long construction lines.
The two short construction lines have been used to create the four short,
single-curved lines on the four tabs.

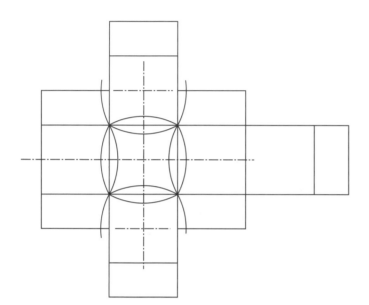

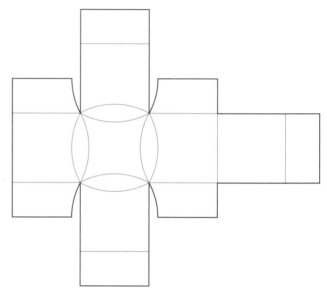

4.9.6 Step 6

This is the completed net. If you are making it by
hand, it will be impossible to make perfect curved
creases. The technique is to draw, and then to
accurately cut out from scrap card, a curved
wedge (like a piece of pie – see the drawing below)
which has the same radius as the curve you wish to
draw. Place this template on the net in the correct
position and create the crease in the normal way,
using the curved edge as the guide for your
creasing implement.

4.10 Single Curves

Whereas double curves (see 4.9) can be placed almost randomly on a net, single curves must be placed in a precise, coordinated pattern. They must be placed on the vertical edges of a box in a mirror-image pattern, and the box must have an even number of these edges (probably four, six or eight – a cube has four). The formula for deciding the radius of the curves is the same as for the double curve: the radius is the length of the edge, plus 5 per cent.

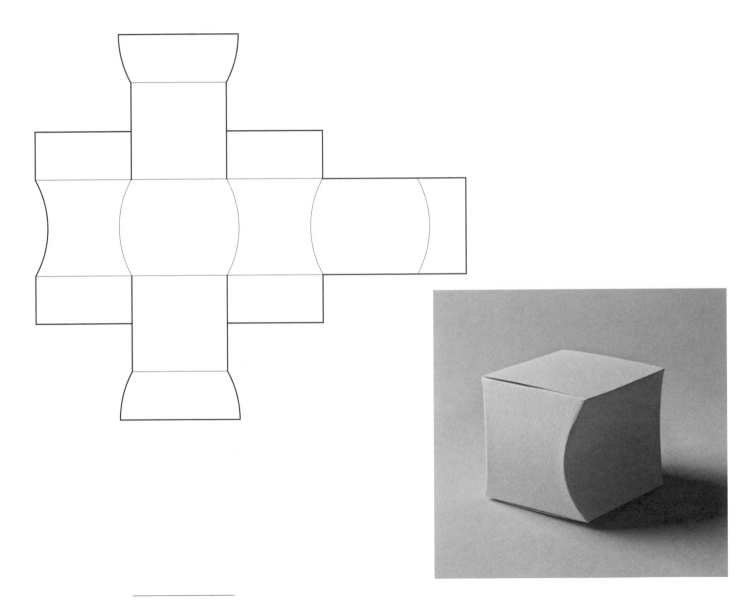

05:

COMMON
CLOSURES

Introduction

The tabbing system described in earlier chapters will hold together strongly almost any straight-edged, flat-faced packaging form (and certain curved-edged, curved-faced forms), but tabs alone will not lock every form tight shut. It is sometimes necessary to add a special lock to the lid or to the base, or both, or to include a glue tab. These extras should not be thought of as compromises, but as important parts of the design, whose inclusion gives strength to an innovative design and so makes it totally practical.

The four most common closures are described on the following pages.

5.1 Glue Tab

Glue gives a package its integrity, holding it together strongly so that it cannot unravel.

Almost every mass-produced example of carton packaging contains at least one glue tab (sometimes called a 'glue line'), if not several. Everything from cereal packets to juice cartons, and from large corrugated boxes containing electrical goods to small fancy gift boxes, will be glued, or alternatively held together along a glue line with industrial staples.

Many times, the glue is applied while the box is collapsed flat so that the glue tab can adhere better to the face to which it bonds. It is much more difficult – and expensive – to apply glue while the package is three-dimensional. This flat-pack system of package design is wonderfully practical, because it means the packages can be stored and transported flat, then quickly erected into three-dimensions to be filled.

However, the problem with flat-packing is that it severely limits the forms that can be created. Few three-dimensional forms can be made as nets and flat-packed. This is one of the reasons why rectangular packages are so common – they can be glued and flat-packed very easily. It is difficult to be innovative with glued, flat-packed packaging.

With lower manufacturing runs, hand assembly becomes an option. Glueing then becomes relatively straightforward and can be done on any tab on any net.

If you are not making packaging, but are designing nets for another purpose, you may not need to use glue at all. Alternatively, you may choose to glue every tab to create a solid, unopenable brick.

An effective alternative to a glue tab is an unglued Tongue Lock (see page 72).

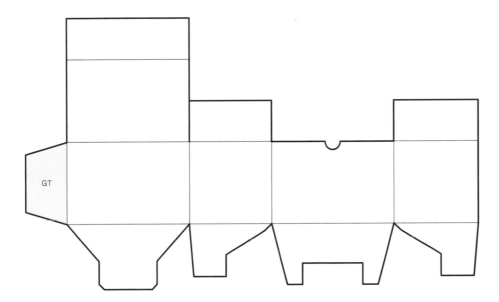

This is the net for a Crash Lock (see page 74). The glue tab is on the left.

Click Lock

The Click Lock is one of the uncelebrated minor miracles of design. It is very simple to make and strengthens enormously the locking of a lid into the body of a box. Such is its strength that, well made from card of just the right weight, it can lock so securely that the lid will open only if the card is ripped!

It is generally used only on smaller packages made from lighter weights of card or board that contain items of no great weight. It can be opened and closed many times without damaging the box, so it is often incorporated on packages that need to be accessed frequently, such as boxes for paper clips.

Although admirably simple in design and use, the Click Lock must be made precisely. Please follow the measurements given below very accurately. The dimensions of the lid and the lid tab are unimportant, but it is crucial to use the 5mm and 2mm measurements given, not to scale them up or down.

The Click Lock may also be used for the base of the package. However, you need to be very sure that when the box is picked up, the lock will be strong enough to prevent the contents falling through.

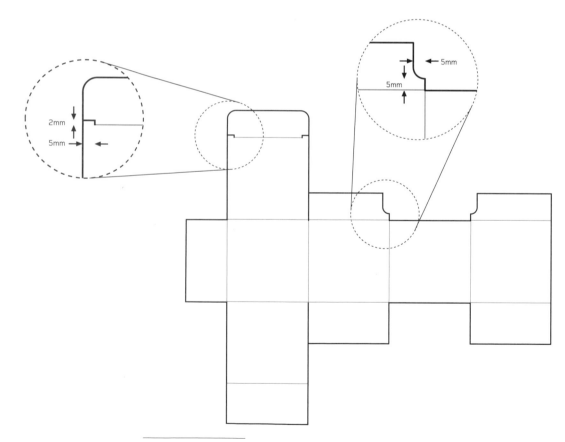

A Click Lock is particularly useful in locking a lid tab – or any other unglued tab – that tapers (that is, a tab with corners of less than 90°). A design consisting of many tapering tabs that tuck into triangular or tapering faces will not lock securely. Adding Click Locks to these non-lid tabs will lock them in.

Tongue Lock
The Tongue Lock is the ultimate in secure, non-glued closures. If you are designing something that may possibly be subjected to a lot of handling, it should be used in preference to the less secure Click Lock. Similarly, it should be used in preference to the Click Lock for larger packages made from corrugated board that will contain heavy items.

It can also be used in place of a glue tab to hold together very securely edges that are not intended to be opened.

For non-packaging designs that you want to look as geometric and as pristine as possible, consider using a Tongue Lock to hold a long, raw edge of the net to the tab beneath, thus straightening an edge which may have a tendency to bow in an unsightly manner. For extra-long edges which need securing, two or more Tongue Locks may be used.

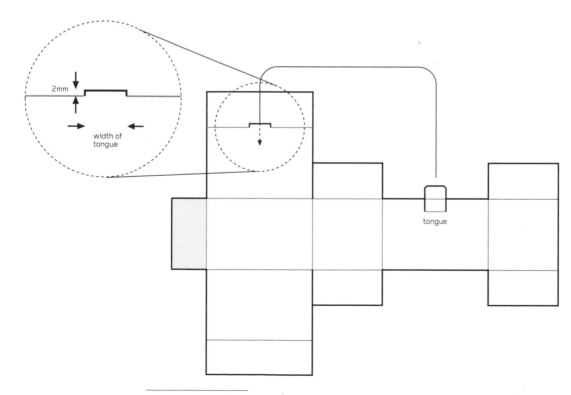

2mm

width of
tongue

tongue

The construction is simple. The tongue and the slit are the same width.
The slit projects forwards into the lid tab by about 2mm. If corrugated card
is used, the slit may need to protrude more depending on the thickness of the
material. The letter-box opening beneath the slit will become apparent only
when the lid tab is folded through 90° (there is no apparent opening when
the card is flat).

A Tongue Lock can be combined with a Click Lock on the same lid.

Crash Lock

Whereas the Click Lock and Tongue Lock are locks to secure a lid, the Crash Lock secures the base of a package. By dividing the base into four sections then interlocking them in a simple but ingenious way, the base is not only secured, but strengthened – a Crash Lock base can hold more weight than a single-layer Click Lock or Tongue Lock base.

The Crash Lock is suitable for any rectangular base, or for any quadrilateral with or without 90° corners. There are interesting variations for five- and six-sided bases.

 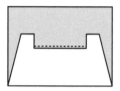 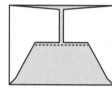 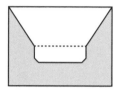

5.4.1
The shaded rectangle is the shape of the base. Draw it carefully. Draw a line parallel to the long edges partway across the middle. Its length should be about 50 per cent of the length of the long edge, perhaps more.

5.4.2
Joined to a long edge, draw the shape shown here. The exact shape and size of the two teeth are unimportant, but it is important to run the horizontal edge just a little below the centre line drawn in 5.4.1.

5.4.3
Joined to the short edges, draw two mirror-image tabs. Note that the horizontal edges are just a little above the centre line drawn in 5.4.1.

5.4.4
Joined to the top edge, draw the shape shown here. Note that the long sloping edges connect the top corners of the base rectangle with the ends of the line drawn in 5.4.1.

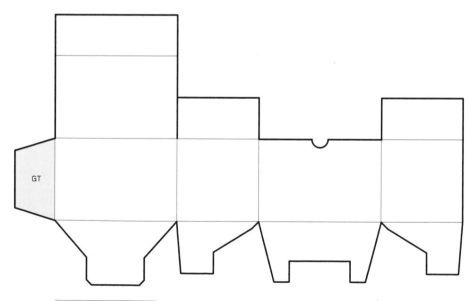

5.4.5
The four tabs can now be drawn on the net of the package, in this configuration. Note that the tab made in 5.4.4 (the tongue) is adjacent to the glue tab at the left-hand edge of the net.

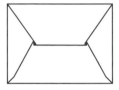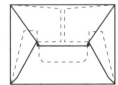

5.4.6
This is how the Crash Lock
looks when the four tabs
are interleaved. First fold
in the tab made in 5.4.2,
then the two smaller
side tabs, then finally the
tongue tab, tucking the
tongue through the slot
across the middle of
the base.

This is the base of the
Crash Lock. To show how
the four sides of the card
interleaf, the length of
the vertical slits each
side of the slot have been
exaggerated. In reality,
they should be tight shut.

06:

CREATING WITH THE SYSTEM

Introduction

This final chapter, also the longest, puts together everything from the preceding chapters to show how to create your own innovative, self-locking forms for packaging or other uses. You will gain most from the chapter if you have already read the rest of the book and understood the principles it presents.

It's often said that 'There's nothing new in geometry'. While it may be true that the well-known 2-D polygons (triangle, pentagon, etc.) and 3-D polyhedra (cube, pyramid, etc.) are well explored individually, they can be combined and deformed in a never-ending series of permutations to create a very great number of beautiful and practical forms. 'Theme and Variation', which starts overleaf, gives strategies for creating these forms.

It is followed by the long 'Creative Examples' section, which explores to the outer limits of what is possible. As always, everything that follows uses the system of net construction described in Chapter 2, though occasionally, to give the forms more strength, the rules are broken.

6.1 Theme and Variation

6.1.1 Single Deformations

Chapter 4 described the basic ways that a solid form can be deformed.
These are:

— Shaving a face
— Shaving an edge
— Shaving a corner
— Twisting a 3-D form
— Compressing or stretching a 3-D form
— Substituting straight edges with curved edges

For each single deformation theme there are numerous variations,
some more significant or beautiful than others.

Shaving a face

Shaving an edge

Shaving a corner

Twisting

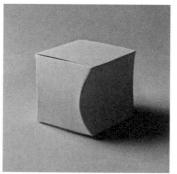

Compressing

Substituting single curved edges

Substituting double curved edges

The four examples on this page show some of the ways in which a corner can
be shaved off a cube. Depending on the angle and extent of the shaving, the
triangular face created can be small, large, equilateral, isosceles or scalene.
The first example reproduces the 'Shaving a Corner' form (see page 56).
The others show a few of the many variations that are possible.

Instead of shaving a single corner, another way of deforming a solid form
may be chosen from the list opposite and subjected to similar variations
of size and angle.

Even the relatively simple idea of deforming a cube only once enables a great
many forms to be designed.

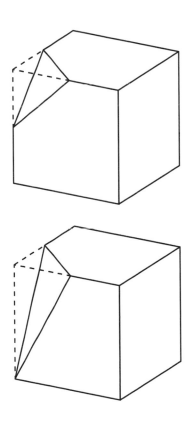

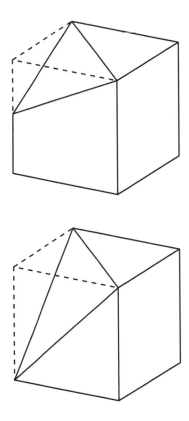

6.1.2 Multiple Deformations

Instead of deforming a cube only once, the same deformation may be performed twice or more.

The examples shown here take the theme of shaving an edge. The first one (1) reproduces the form on page 55, shaving just one edge off the cube. The second example (2) shaves off two opposite edges from the top face of the cube to create a form reminiscent of a gable-roofed house. The third example (3) shaves off two adjacent edges from the top face of the cube to leave only a small square face on the top. The fourth example (4) shaves three edges off the top face to create a hip roof. It is possible to make many other forms by shaving off upper, side and lower edges in a variety of combinations – astonishingly, about 150 can be created (excluding mirror image and rotational repeats) just by shaving off edges at 45°, as shown here! Suddenly, it becomes easy to make forms that no one has seen before. That's quite a thought.

Other deformations from the list on page 78 may also be performed more than once, especially the shaved and curved options.

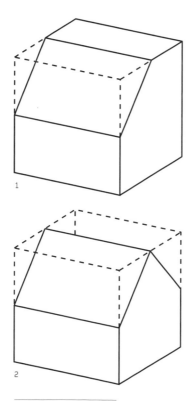

1

2

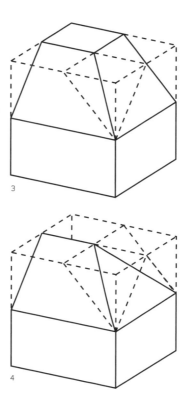

3

4

6.1.3 Combinational Deformations

Deformations can be performed in combination almost to infinity. Many of the combinations can be tawdry, weak, ugly or plainly ridiculous. Your job as a designer is not to find them all (life is too short), but to find the ones that work for you; to limit your options ... put simply: to design.

The deformations need not be performed only on a cube. It is possible to deform cuboids, pyramids (three-, four-, five-sided, etc.), prisms (triangular, pentagonal, hexagonal, etc.) and a great many other basic solids.

Here are a few suggestions from the almost limitless options available to you. The first four are illustrated on this page:

— A square cuboid, twisted with a double curve.
— A truncated pyramid with a shaved face
— A tall hexagonal prism with three shaved corners
— The same, but a squat hexagonal prism

and also...
— An asymmetric five-sided pyramid
— A pentagonal prism with a shaved pentagonal face
— A cuboid with two small double curves on one edge
— A hexagonal prism with C curves down the sides
— A truncated octagonal pyramid
— A cube with a double curve on every edge
— An asymmetrically faceted pentagonal prism
— Two pyramids joined as one net, base to base
— A faceted, truncated, triangular pyramid
— A cuboid, with one corner shaved off in such a way that it can stand and balance on the shaved triangular face
— A cube with a lid occupying two square faces ... and so on and so on, seemingly without end!

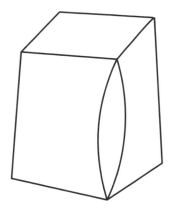

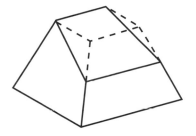

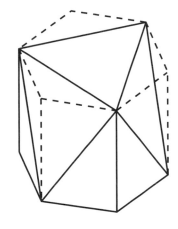

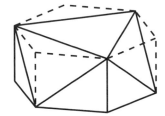

6.2 Creative Examples

The examples that follow were created by second- and third-year degree-level students of communication design and product design at the Hochschule für Gestaltung, Schwäbish Gmünd, Germany, over four consecutive days. The first two days were a workshop, in which the students as a group followed the steps in Chapter 2 to learn how to make a net, after which they made many of the Deforming a Cube examples in Chapter 4. This was followed by a two-day project, in which the students could make what they wanted, but with some guidance regarding which form to make and how to make it. Throughout, only basic geometry equipment was used, though towards the end a few students used 3-D CAD software to help calculate angles and edge lengths that were very difficult to calculate accurately with pen and paper.

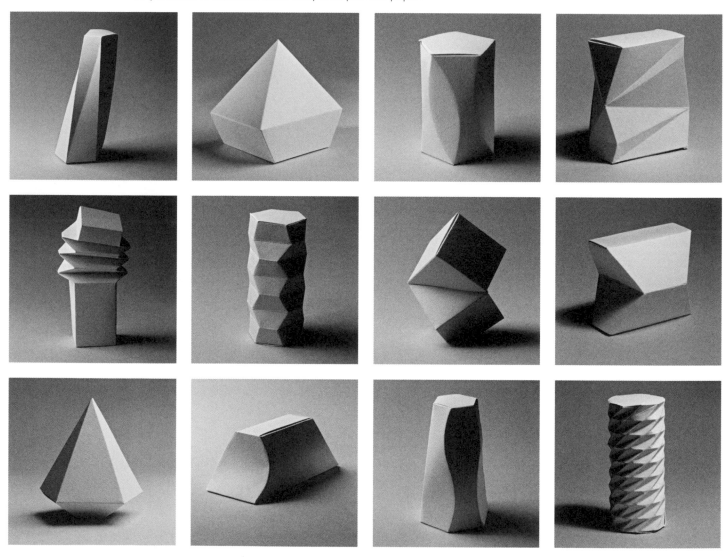

After the project was finished, the students drew nets of their designs and emailed them to the author, who used a plotter to recreate the designs in card. It is these recreations that can be seen in the photographs that follow. The net drawings are by the students, with only occasional modifications.

The project is typical of many run by the author. Students who were unaccustomed to working three-dimensionally, or were nervous when dealing with problems that involved a little mathematics, soon began to create interesting work with confidence and enthusiasm. Those with more confidence often had to be restrained from making designs that were too complex, incorporating too many types of deformation in one form. Many students submitted net drawings of two or more very different forms.

After making such playful forms, it will be a relatively simple matter for the students to rein in their creativity in order to solve real-world design problems – whether packaging or otherwise – using the system of net construction presented in this book.

What they achieved, you can also achieve.

   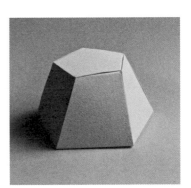

6.2.1

This sophisticated and subtle form is an octagonal prism which has been
twisted (see 'Twisting Opposite Faces', page 59) then had its top edge shaved
off at an angle (see 'Shaving a Face', page 54). Its bottom face is a regular
octagon, but the top one – as can be seen from the drawing – is irregular.
The irregularity is caused by the distortions to the edges made by the
twisted faces joining the shaved face.

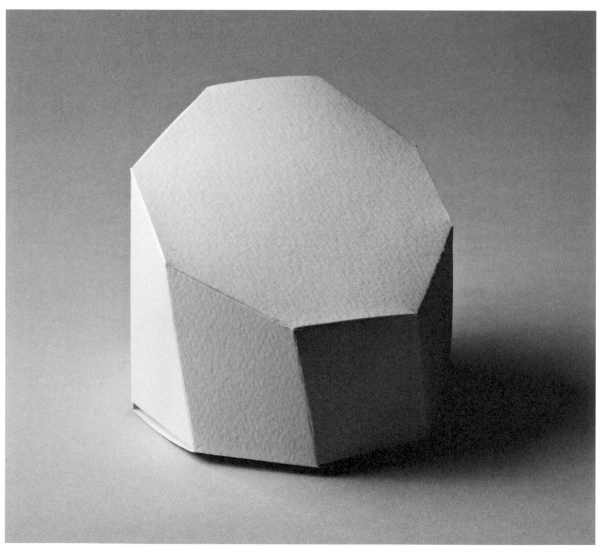

6.   CREATING
     WITH THE
     SYSTEM
6.2 Creative
     Examples

6.2.1

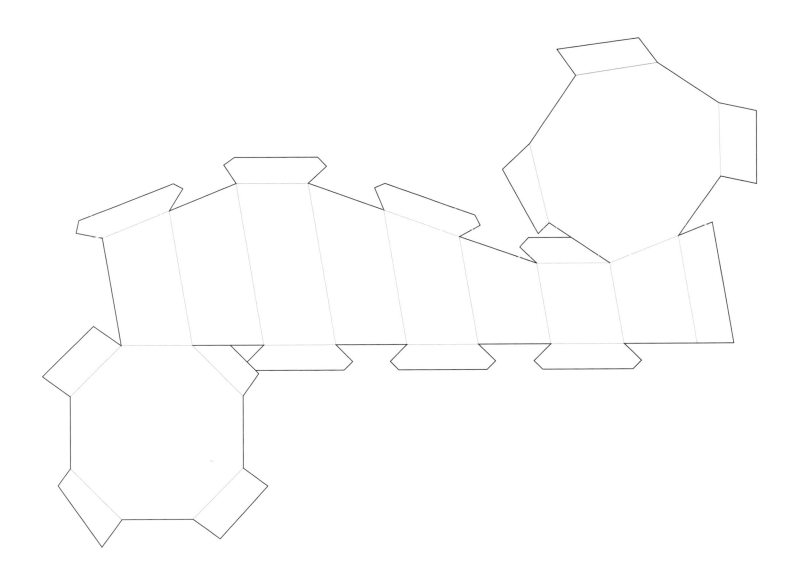

6.	CREATING
	WITH THE
	SYSTEM
6.2	**Creative**
	Examples
6.2.2	

6.2.2

Two pentagonal pyramids meet around the widest point, one taller and more slender than the other. To help hold the long, vertical edge of the taller pyramid tight shut, the tab has been divided into three. The clever inversion of the bottom corner allows the form to stand when it would otherwise have had to lie forlornly on its side. Almost any corner on any net can be inverted in this manner, allowing a form to then stand upright on that corner.

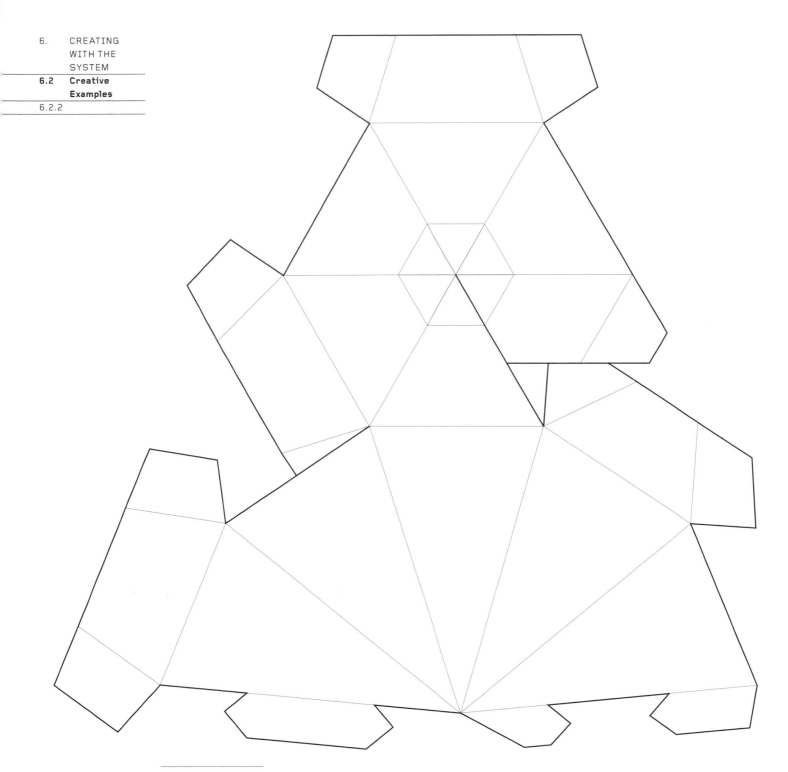

6.2.3

The gentle curves used here create a form considerably more graceful than the prosaic straight-sided truncated rectangular pyramid from which it is derived. All the tabs are 90° or greater, ensuring that they lock the form together with maximum efficiency and strength. The single curve technique (see 'Single Curves', page 65) is here used to great effect.

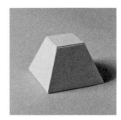

6. CREATING
WITH THE
SYSTEM

**6.2 Creative
Examples**

6.2.3

6. CREATING
    WITH THE
    SYSTEM
6.2 **Creative
    Examples**

6.2.4

6.2.4

Beneath the complex faceting lies a very conventional cuboid box. The vertical edges have been twisted off the vertical, though, interestingly, they always remain parallel. A pair of opposite corners within each resulting quadrilateral is then connected with an extra fold, dividing the surface into a seemingly random array of triangles. However, close study reveals that the surface is highly structured and any apparent randomness is illusory, caused by the absence of any 90° corners.

6.   CREATING
     WITH THE
     SYSTEM

6.2 Creative
     Examples

6.2.4

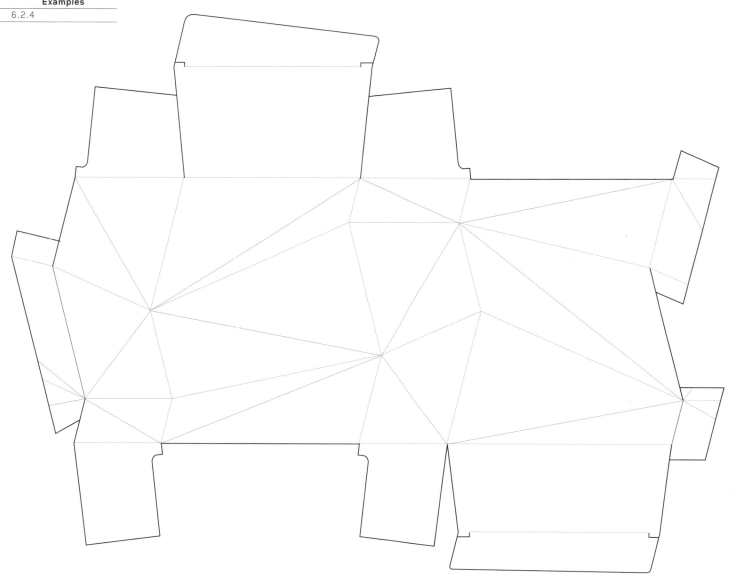

6.2.5

Complex geometry allows this truncated pentagonal pyramid to both twist and lean. Additional valley folds divide the tall, twisting quadrilaterals into triangular facets, allowing this otherwise stressed structure a high degree of stability. If it was taller, it would tip over; if it was shorter, the leaning effect would be less dramatic.

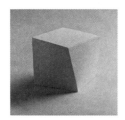

6.   CREATING
      WITH THE
      SYSTEM

6.2   Creative
      Examples

6.2.5

6.   CREATING
        WITH THE
        SYSTEM

6.2   **Creative**
        Examples

6.2.6

6.2.6
A conventional rendering of this form would have the S curves in opposition
from one edge to the next around all six sides. However, every third edge
is straight, interrupting the flow of the curves around the box and creating
a striking contrast between the straight and curved edges. Note how one
straight edge is 'zipped up' by a series of small tabs. This helps to create
a strong, flat, straight edge, in contrast to the curves.

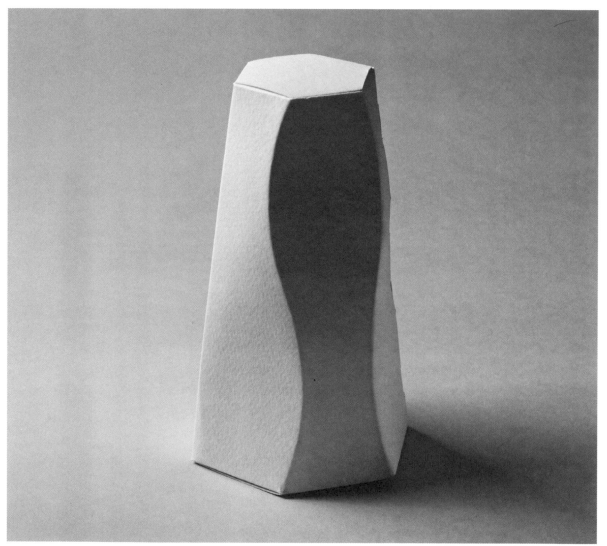

6.   CREATING
      WITH THE
      SYSTEM
6.2   Creative
      Examples
6.2.6

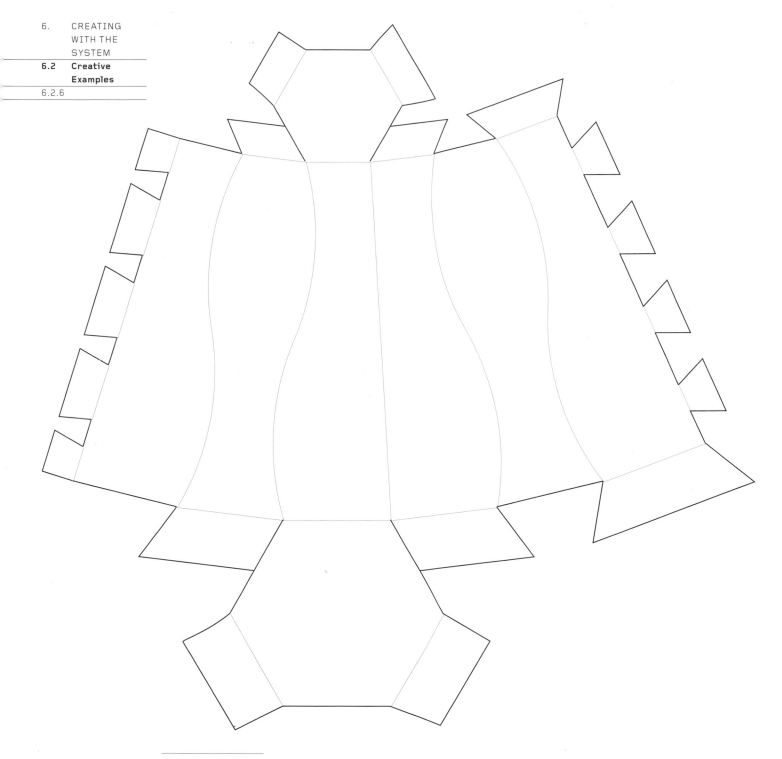

6.2.7

This droll design is a stack of 'Compressing Face-to-face' examples (see page 61) atop a conventional A x A x B cuboid box (see 'Square Cuboid Boxes', page 42). A glue line is necessary to hold the form together, because tabbing would be too complex and too imprecise to be effective down the ladder of horizontal folds. Note how the overlapping of the card at the glue line is done not along a vertical edge, but down the centre of a face. This solution is considerably simpler than cutting a long zigzag down the edge of the card at the glue line.

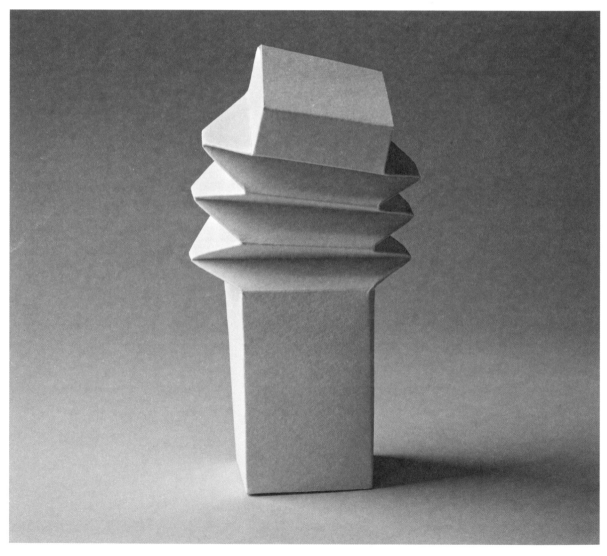

6. CREATING
WITH THE
SYSTEM
6.2 **Creative**
Examples
6.2.7

6.2.8

Comparisons between this piece and 6.2.7 are instructive. Like 6.2.7, the design is a stack of 'Compressing Face-to-face' examples (see page 61), though with a hexagonal cross-section. However, unlike 6.2.7, the angles of the zigzag mean that the form cannot be squeezed flat. In this piece, the zigzag edge at the side of the net is zipped up with a series of tabs. It also has a Click Lock lid (see page 70).

6. CREATING
   WITH THE
   SYSTEM

**6.2 Creative
   Examples**

6.2.8

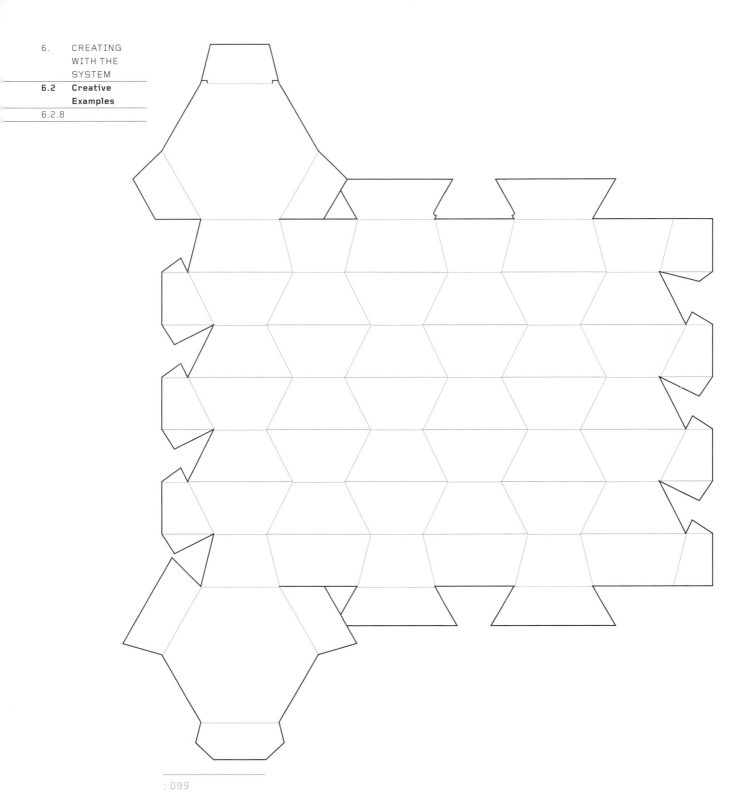

6.   CREATING
      WITH THE
      SYSTEM
6.2   Creative
      Examples
6.2.9

6.2.9

This fascinating structure is created entirely from edges and folds that are horizontal, vertical or 45°. There are a great many possible configurations for the net. The one chosen maximizes strength at the three inverted (valley fold) edges around the waist. Note how those edges are locked together with half-length tabs which hook around each other and how there is no fold line where these tabs join the visible face. The inverted corner permits it to stand upright (see also 6.2.2).

6.   CREATING
      WITH THE
      SYSTEM
6.2   Creative
      Examples

6.2.9

6. CREATING
WITH THE
SYSTEM
**6.2 Creative
Examples**
6.2.10

6.2.10

The form is an example of 'Compressing Face-to-face' (see 'Deforming a
Cube', page 61), but made with an A x A x B cuboid (see 'Square Cuboid Boxes'
page 42) instead of a cube. However, when the box is folded up, the A x A
(square) silhouette will reduce in height, so that the box appears to be an A
x B x C cuboid. The student chose to connect the narrow, rectangular face at
the bottom to the net by a short edge instead of a long edge, in contradiction
to the advice given in Step 6, Chapter 2 (see page 21). If the face had been
connected to the net by a long edge, it would have been locked in with two small
wedge-shaped tabs joined to the short edges, thus making it more difficult to
close than the net shown here. So although this net looks incorrect, it is easier
to open and close than the correct net.

6. CREATING
WITH THE
SYSTEM

6.2 Creative
Examples

6.2.10

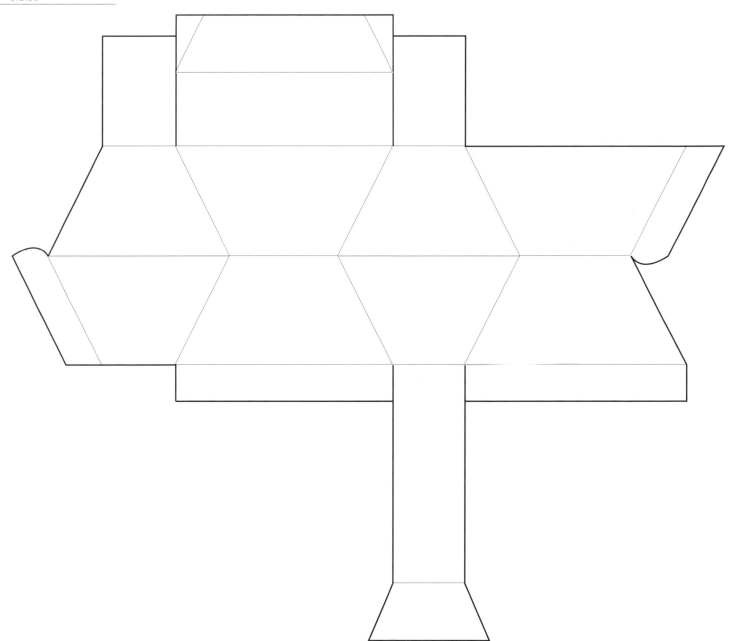

6.   CREATING
     WITH THE
     SYSTEM

6.2   **Creative**
     Examples

6.2.11

6.2.11

The subtle twist of this truncated pentagonal pyramid creates a complex and enigmatic form. It is a combination of the Truncated Pyramid constructed step-by-step in Chapter 2 and 'Twisting Opposite Faces' (see page 59). The many unusual angles mean that the tabs have to be designed with particular care. Despite the distortions around the sides of the form, the top and bottom faces are regular pentagons.

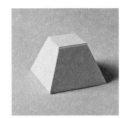

6.   CREATING
       WITH THE
       SYSTEM

6.2   **Creative**
       Examples

6.2.11

6.2.12
Here, a simple square base is topped by a non-square (rectangular) pyramid, transforming an otherwise routine form into something subtle and unconventional. Note the three flange tabs (see 'Tapering Tabs', page 36) which are vital for locking loose, tapering tabs tightly into the assembled form.

6.   CREATING
     WITH THE
     SYSTEM

6.2   Creative
     Examples

6.2.12

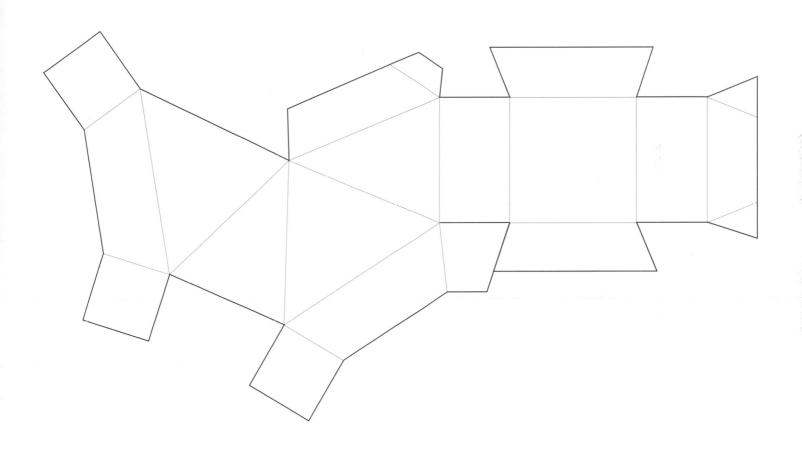

6. CREATING
WITH THE
SYSTEM

6.2 Creative
Examples

6.2.13

6.2.13

This is the only photograph in this section to show forms fitting together. It is a combination of the truncated pyramid constructed step-by-step in Chapter 2 and 'Twisting: Faceted Version' (see page 60). However, there is one subtlety that makes the forms fit snugly together: the top square is half the area of the bottom square (the diagonal of the small square is the same length as the side of the large square). With squares of this proportion and a form of any height, four of the eight triangles will rise vertically and fit against triangular faces on other forms which also rise vertically. Thus, the form will tessellate infinitely in two dimensions and stack in three dimensions.

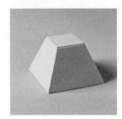

6.   CREATING
WITH THE
SYSTEM

**6.2   Creative
Examples**

6.2.13

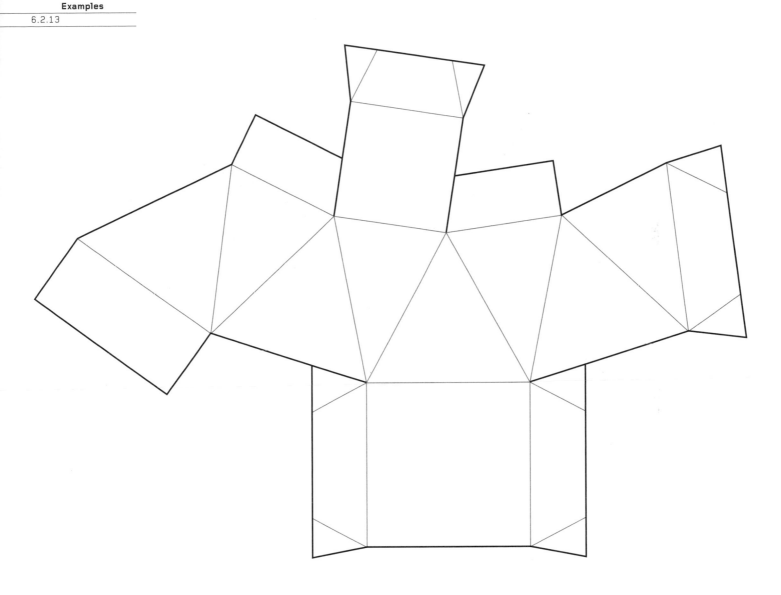

6.	CREATING
	WITH THE
	SYSTEM
6.2	**Creative**
	Examples
6.2.14	

6.2.14

Sometimes, less is more. Here, the horizontal valley-fold waist is left unfolded, so that the top and bottom halves of the form connect along a curved edge, adding beauty to an already interesting form. Note how a simple, straight glue line closes the form, thus diverting the edge of the card away from the folds so that they may have maximum impact.

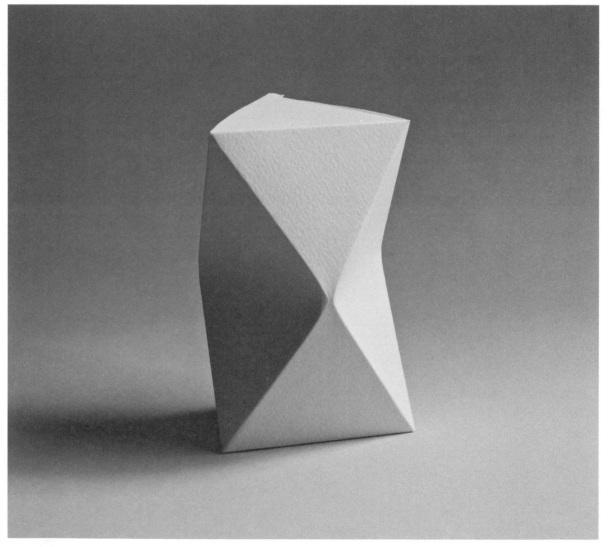

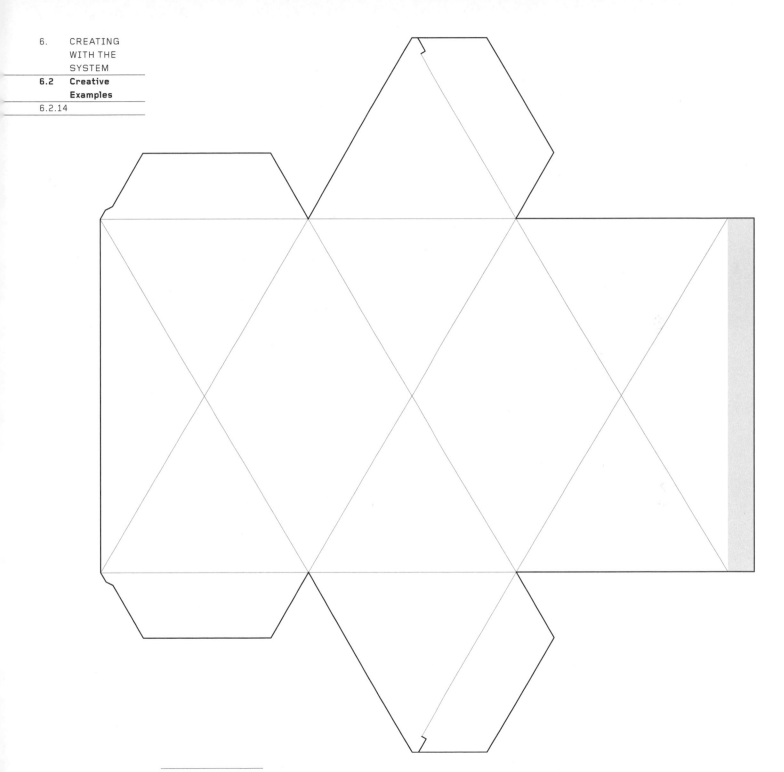

6.2.15

The form looks like an offcut, as though it was cut away from something more substantial ... and that's just what it is. If the two small triangles were removed and the faces continued, the form would resemble the corner cut off the cube in 'Shaving a Corner' (see page 56). By further shaving off the two sharp corners, the form becomes more compact.

6.   CREATING
      WITH THE
      SYSTEM
6.2   Creative
      Examples
6.2.15

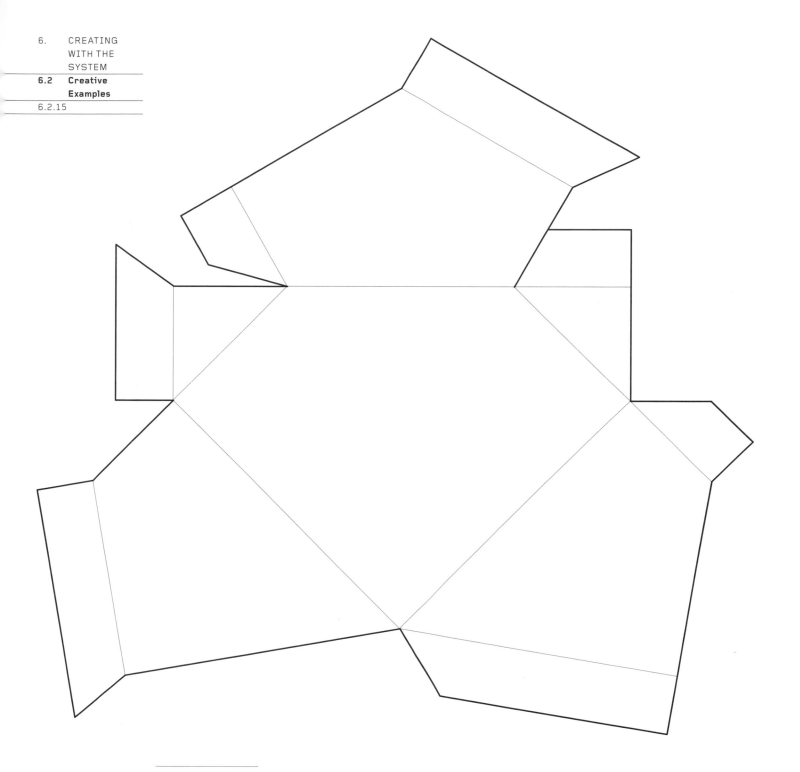

6.2.16

Beneath the swirling lines lies a combination of the two forms constructed step-by-step in Chapter 2 (a truncated pyramid and a hexagonal prism), combined with 'Double Curves' and 'Single Curves' (see pages 62 and 65). The result is a form which bulges and curves everywhere and which seems not to have been made from a single sheet of card.

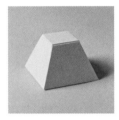

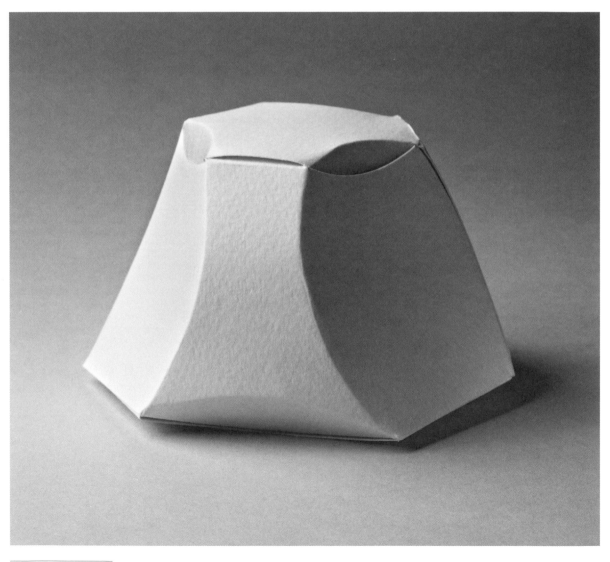

6.   CREATING
      WITH THE
      SYSTEM

**6.2   Creative
         Examples**

6.2.16

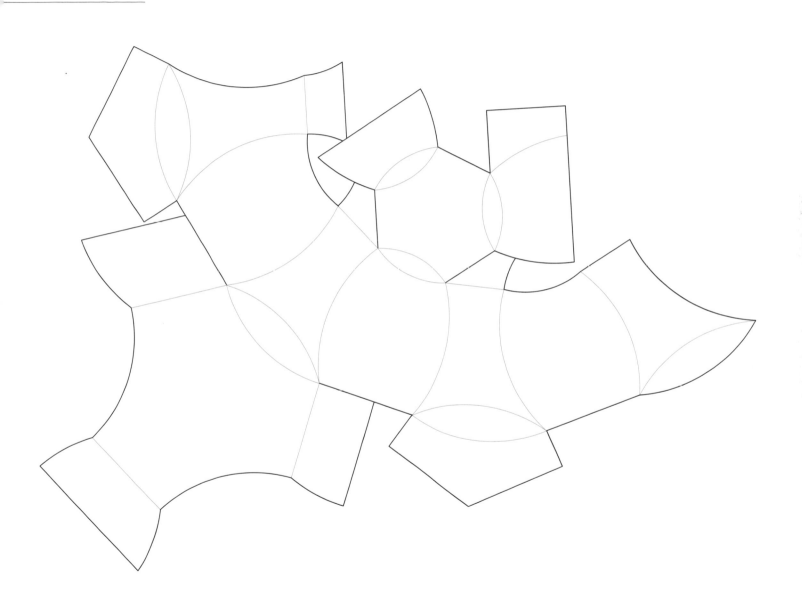

6.2.17

Not for the faint-hearted, this stunning structure shows the extremes to which the system of net construction explained in this book can be pushed, yet still work well. Although highly complex to draw and make, the complexity comes from repetition – in fact, there are very few different elements in the design. Many of the designs in the book could achieve this level of complexity if some of their elements were repeated many times, as here. It is at this point that boxes or packages become sculpture.

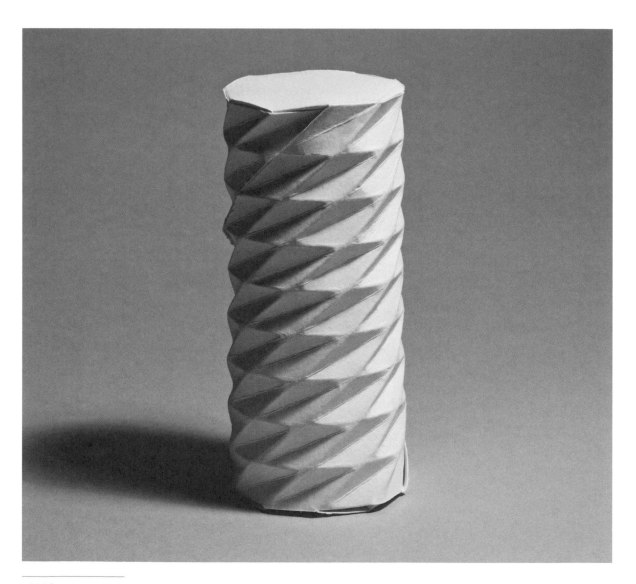

6.   CREATING
       WITH THE
       SYSTEM

6.2   Creative
       Examples

6.2.17

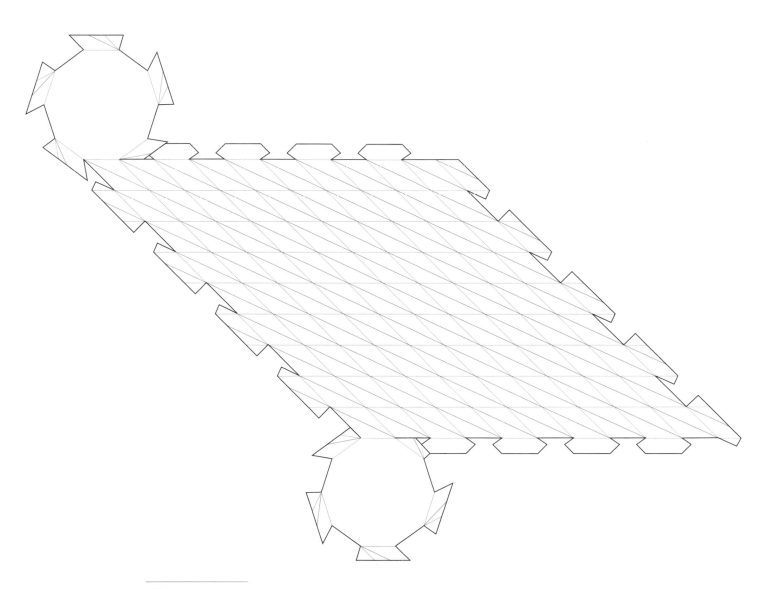

6.2.18

A pentagonal prism is twisted (see 'Twisting Opposite Faces', page 59) and double curves replace straight edges on four of its vertical edges (see 'Double Curves', page 62). The result is an elegant form, given added interest by the straight edge. Double curves work well on square and pentagonal forms, but less well when the number of sides increases, because they become increasingly flat.

6.   CREATING
       WITH THE
       SYSTEM

6.2   Creative
       Examples

6.2.18

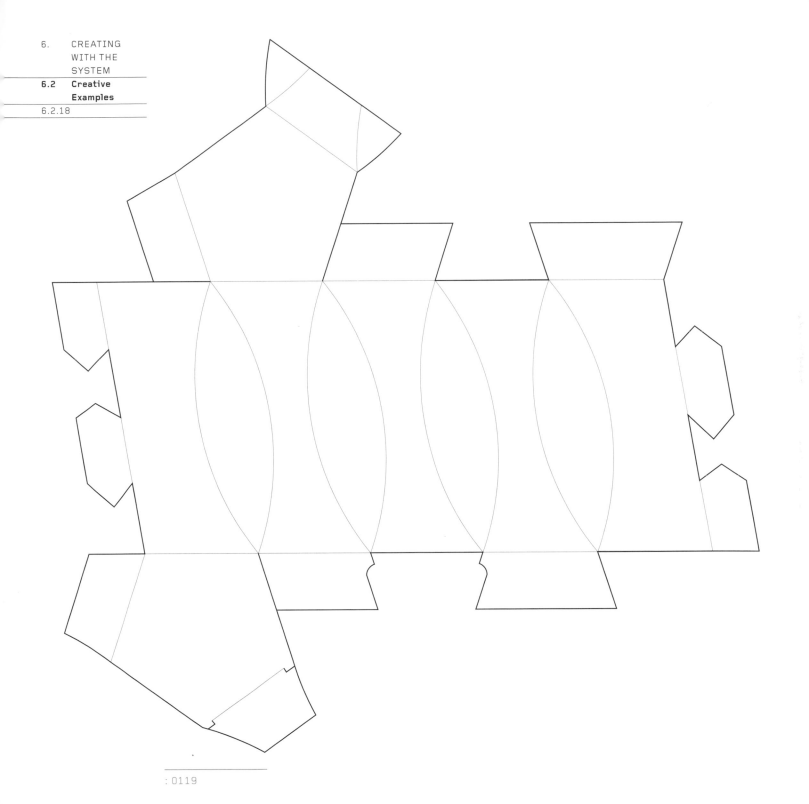

6.2.19

A conventional pentagonal truncated pyramid (on the left) has an additional valley fold added across each of its vertical faces. When all five mountain and all five valley folds are twisted simultaneously, the form collapses and locks to create the form on the right. Although complex on the outside, inside the spaces are clean and simple – they are pentagonal pyramids, the top one being upside down. The same twisting can be performed on other prisms and truncated pyramids, though care needs to be taken if they are unusually tall, as they do not twist into shape well.

6.   CREATING
       WITH THE
       SYSTEM

6.2   **Creative**
       Examples

6.2.19

6.   CREATING
      WITH THE
      SYSTEM
6.2   Creative
      Examples
6.2.20

6.2.20

This is the most minimal net in the book. All the folds around the sides of the form have been removed, so that everything curves softly. Note how it has a hexagonal base, but a triangular top – six edges at the bottom become three at the top, creating three triangles and three trapeziums around the vertical sides. The three-piece interlocking lid ensures that the top is simple, clean and symmetrical, adding greatly to the successful minimal aesthetic of the form.
On many nets, many folds can be removed to create a series of elegant curves.

6.   CREATING
      WITH THE
      SYSTEM

6.2   Creative
      Examples

6.2.20

6. CREATING
    WITH THE
    SYSTEM

6.2 **Creative**
     Examples

6.2.21

6.2.21
Instead of locking all tabs inside a three-dimensional form, some of them – or indeed, all of them – can be locked on the outside. This example shows how one tab is locked on the outside by being tucked into a slit. Such a lock signals that the tab is the lid and that releasing it will open the box

6.   CREATING
      WITH THE
      SYSTEM
6.2   Creative
      Examples

6.2.21

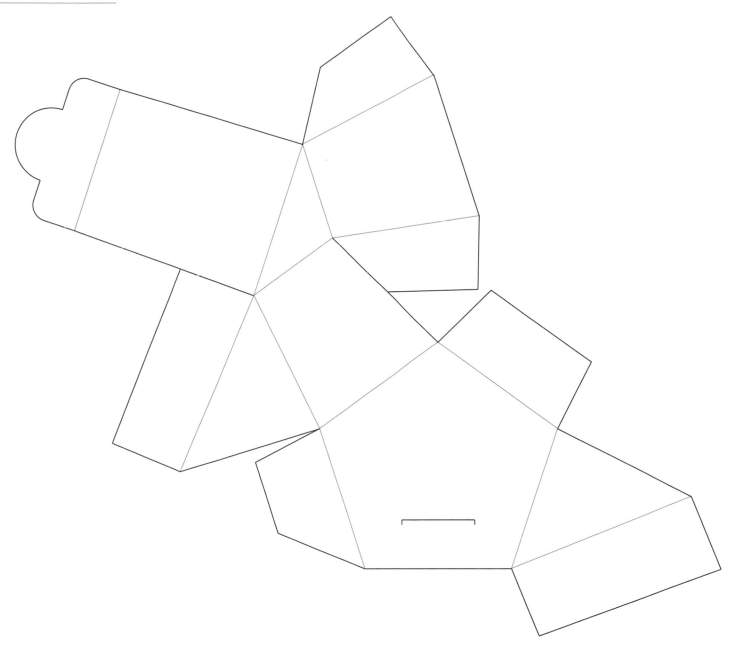

How Do I Produce My Box?

By now, you will have used the method in this book to create a prototype of a design. After the prototype has been approved by everyone who needs to give their opinion, it can be made. Depending on the material used and the number of copies you wish to make, there are several options open to you.

1. Using a Computer Printer

For low production runs using card it is possible to print out the net on an A4 or A3 computer printer, then cut and fold each design by hand. This is time-consuming, exacting work, but significantly less expensive than any manufacturing option. A printer will also, of course, print any surface graphics you may wish to include.

If your printer bends the card through 180° as it prints, it may not be able to handle card heavier than 200gsm. The solution is either to invest in a printer that passes paper and card straight through without bending it, or use a friend's straight-feed printer. If no suitable printer is available, use the Pricking Method, described below.

2. The Pricking Method

For low production runs using heavy boards, corrugated card, plastics and other materials which are either too big or too heavy to pass through a computer printer, draw the net accurately on paper and tape the paper to the relevant material. Then make a pinprick through the drawing at each corner of the net into the material beneath. Remove the drawing, then connect the pin holes as in a child's 'join the dots' puzzle, to recreate an accurate copy of the original net. The same paper template can be used many times, providing the pin holes remain small. This low-tech method of reproduction may seem primitive, but it is extremely effective.

3. Plotters and Laser Cutters

Though still specialist technologies, flat-bed plotters and laser cutters are increasingly commonplace. Almost all packaging companies have a plotter in which the drawing pen is replaced with a knife, which is used to test new designs and create prototypes for clients. Plotters are amazing, addictive toys! Some companies will hire theirs out to people who want to make a low production run. If your design is difficult to make by hand and the production run is too low to justify manufacturing it, consider negotiating with a packaging company for the use of their plotter.

Laser cutters are also useful for low production runs, but come into their own for cutting intricate detail into card. The level of detail they can create is truly extraordinary. The drawbacks with laser cutters are that they can be expensive to use and they can leave unsightly brown scorch marks along cut edges.

4. Professional Production

For longer production runs, your design will need to be manufactured.

The first step is to contact local packaging companies and show them your design. Be prepared to be given advice about how your hard-won design can be improved for manufacture: how some of the folds must be moved by fractions of a millimetre to accommodate the thickness of the card when it is folded through 90°; how tabs must be made longer or shorter, or be given rounded corners; how a glue line can (or must) be added; how the security of the lid closure can be improved; which card or board should be used; how to print the surface graphics; how to cut costs ... and a dozen other items of good advice born of their professional experience. This book would need to triple in length to include all these finesses — and then, to be frank, a professional packaging engineer would change them all anyway to suit his/her design idiosyncrasies and the specifications of the machinery used at the production plant. What you have learnt from this book is more than enough for you to take your design to a professional packaging engineer for completion.

It is well worth approaching several packaging companies for advice. Each will have its own way of improving a design for manufacture. Some will have in-house printing facilities and some will welcome innovation, whereas others will baulk at the thought of producing anything other than a square box from an existing template.

Acknowledgements

The author would like to thank the many art and design colleges who, over many years, have allowed him to work with their students to refine the system of net construction presented in this book, and also to thank the many students who participated.

Extra-special thanks go to Gilad Barkan (CEO) and Behnaz Shamian-Hershkovitz (Designer) of Gilad Dies Ltd in Holon, Israel (barkang@netvision. net.il: Tel+972-3-5583728) for their dedication and expert help processing many net drawings through their plotter. You were amazing and made this book possible.

Thanks also to Professor Emma Frigerio of Milan University for valuable help with some of the mathematics. My thanks also to the Rector, Cristina Salerno, and Professor Peter Stebbing of the Hochschule für Gestaltung, Schwäbisch Gmünd, Germany, for permitting me to work with their students to create the work seen in the final chapter, and to the students themselves, who worked with such dedication and diligence. They are:

Marion Bruells, Christiane Frommelt, Janine Gehl, Thomas Grikschas, Andreas Hogh, Julian Hölzer, Adrian Jehle, Patrick Klingebiel, Moritz Koehn, Juliane Lanig, Bernhard Meyer, Jan Michalski, Katja Mollik, Linda Moser, Christina Müller, Stefanie Nagel, Christine Putz, Olga Rau, Janina Reinhard, Julius Renz, Robin Ritter, Andrea Schmaderer, Sascha Benjamin Simeth, Hakon Ullrich and Anna Kubelik.

WITHD   N
FOR R

New College Nottingham
Learning Centres